Weakly Connected Nonlinear Systems

Boundedness and Stability of Motion

PURE AND APPLIED MATHEMATICS

A Program of Monographs, Textbooks, and Lecture Notes

MONOGRAPHS AND TEXTBOOKS IN PURE AND APPLIED MATHEMATICS

Recent Titles

Kevin J. Hastings, Introduction to the Mathematics of Operations Research with *Mathematica®*, Second Edition (2006)

Robert Carlson, A Concrete Introduction to Real Analysis (2006)

John Dauns and Yiqiang Zhou, Classes of Modules (2006)

N. K. Govil, H. N. Mhaskar, Ram N. Mohapatra, Zuhair Nashed, and J. Szabados, Frontiers in Interpolation and Approximation (2006)

Luca Lorenzi and Marcello Bertoldi, Analytical Methods for Markov Semigroups (2006)

M. A. Al-Gwaiz and S. A. Elsanousi, Elements of Real Analysis (2006)

Theodore G. Faticoni, Direct Sum Decompositions of Torsion-Free Finite Rank Groups (2007)

R. Sivaramakrishnan, Certain Number-Theoretic Episodes in Algebra (2006)

Aderemi Kuku, Representation Theory and Higher Algebraic K-Theory (2006)

Robert Piziak and P. L. Odell, Matrix Theory: From Generalized Inverses to Jordan Form (2007)

Norman L. Johnson, Vikram Jha, and Mauro Biliotti, Handbook of Finite Translation Planes (2007)

Lieven Le Bruyn, Noncommutative Geometry and Cayley-smooth Orders (2008)

Fritz Schwarz, Algorithmic Lie Theory for Solving Ordinary Differential Equations (2008)

Jane Cronin, Ordinary Differential Equations: Introduction and Qualitative Theory, Third Edition (2008)

Su Gao, Invariant Descriptive Set Theory (2009)

Christopher Apelian and Steve Surace, Real and Complex Analysis (2010)

Norman L. Johnson, Combinatorics of Spreads and Parallelisms (2010)

Lawrence Narici and Edward Beckenstein, Topological Vector Spaces, Second Edition (2010)

Moshe Sniedovich, Dynamic Programming: Foundations and Principles, Second Edition (2010)

Drumi D. Bainov and Snezhana G. Hristova, Differential Equations with Maxima (2011)

Willi Freeden, Metaharmonic Lattice Point Theory (2011)

Murray R. Bremner, Lattice Basis Reduction: An Introduction to the LLL Algorithm and Its Applications (2011)

Clifford Bergman, Universal Algebra: Fundamentals and Selected Topics (2011)

A. A. Martynyuk and Yu. A. Martynyuk-Chernienko, Uncertain Dynamical Systems: Stability and Motion Control (2012)

Washek F. Pfeffer, The Divergence Theorem and Sets of Finite Perimeter (2012)

Willi Freeden and Christian Gerhards, Geomathematically Oriented Potential Theory (2013)

Anatoly Martynyuk, Larisa Chernetskaya, and Vladislav Martynyuk, Weakly Connected Nonlinear Systems: Boundedness and Stability of Motion (2013)

Weakly Connected Nonlinear Systems

Boundedness and Stability of Motion

Anatoly Martynyuk

Larisa Chernetskaya

Vladislav Martynyuk

CRC Press
Taylor & Francis Group
Boca Raton London New York

CRC Press is an imprint of the
Taylor & Francis Group, an **informa** business
A CHAPMAN & HALL BOOK

CRC Press
Taylor & Francis Group
6000 Broken Sound Parkway NW, Suite 300
Boca Raton, FL 33487-2742

First issued in paperback 2019

© 2013 by Taylor & Francis Group, LLC
CRC Press is an imprint of Taylor & Francis Group, an Informa business

No claim to original U.S. Government works

ISBN-13: 978-1-4665-7086-3 (hbk)
ISBN-13: 978-0-367-38063-2 (pbk)

Visit the Taylor & Francis Web site at
http://www.taylorandfrancis.com

and the CRC Press Web site at
http://www.crcpress.com

Contents

Preface   xi

Acknowledgments   xv

1 Preliminaries   **1**
- 1.1 Introductory Remarks 1
- 1.2 Fundamental Inequalities 2
  - 1.2.1 Gronwall type inequalities 2
  - 1.2.2 Bihari type inequalities 7
  - 1.2.3 Differential inequalities 12
  - 1.2.4 Integral inequalities 16
- 1.3 Stability in the Sense of Lyapunov 17
  - 1.3.1 Lyapunov functions 17
  - 1.3.2 Stability theorems 21
- 1.4 Comparison Principle 29
- 1.5 Stability of Systems with a Small Parameter 32
  - 1.5.1 States of equilibrium 33
  - 1.5.2 Definitions of stability 34
- 1.6 Comments and References 35

2 Analysis of the Boundedness of Motion   **37**
- 2.1 Introductory Remarks 37
- 2.2 Statement of the Problem 38
- 2.3 μ-Boundedness with Respect to Two Measures 40
- 2.4 Boundedness and the Comparison Technique 47
  - 2.4.1 Auxiliary results 47
  - 2.4.2 Conditions for the boundedness of motion 48
- 2.5 Boundedness with Respect to a Part of Variables 56
- 2.6 Algebraic Conditions of μ-Boundedness 61
- 2.7 Applications 66
  - 2.7.1 Lienard oscillator 66
  - 2.7.2 Connected systems of Lurie–Postnikov equations . . . 67
  - 2.7.3 A nonlinear system with weak linear connections . . . 69
- 2.8 Comments and References 71

3 Analysis of the Stability of Motion   73

  3.1 Introductory Remarks 73
  3.2 Statement of the Problem 74
  3.3 Stability with Respect to Two Measures 76
  3.4 Equistability Via Scalar Comparison Equations 86
  3.5 Dynamic Behavior of an Individual Subsystem 90
  3.6 Asymptotic Behavior 95
      3.6.1 Uniform asymptotic stability 95
      3.6.2 The global uniform asymptotic stability 98
      3.6.3 Exponential stability 99
      3.6.4 Instability and full instability 103
  3.7 Polystability of Motion 105
      3.7.1 General problem of polystability 105
      3.7.2 Polystability of the system with two subsystems . . . 106
  3.8 Applications 109
      3.8.1 Analysis of longitudinal motion of an aeroplane 109
      3.8.2 Indirect control of systems 112
      3.8.3 Control system with an unstable free subsystem . . . 114
  3.9 Comments and References 116

4 Stability of Weakly Perturbed Systems   119

  4.1 Introductory Remarks 119
  4.2 Averaging and Stability 120
      4.2.1 Problem and auxiliary results 120
      4.2.2 Conditions for stability 122
      4.2.3 Conditions of instability 126
      4.2.4 Conditions for asymptotic stability 131
  4.3 Stability on a Finite Time Interval 135
  4.4 Methods of Application of Auxiliary Systems 141
      4.4.1 Development of limiting system method 141
      4.4.2 Stability on time-dependent sets 146
  4.5 Systems with Nonasymptotically Stable Subsystems 151
  4.6 Stability with Respect to a Part of Variables 163
  4.7 Applications 166
      4.7.1 Analysis of two weakly connected oscillators 166
      4.7.2 System of n oscillators 171
  4.8 Comments and References 177

5 Stability of Systems in Banach Spaces   179

  5.1 Introductory Remarks 179
  5.2 Preliminary Results 179
  5.3 Statement of the Problem 181
  5.4 Generalized Direct Lyapunov Method 182
  5.5 μ-Stability of Motion of Weakly Connected Systems 185
  5.6 Stability Analysis of a Two-Component System 196

5.7   Comments and References   200

Bibliography   **203**

Index   **211**

Preface

The investigation of nonlinear systems with a small parameter is attributable to a lot of modern problems of mechanics, physics, hydrodynamics, electrodynamics of charge-particle beams, space technology, astrodynamics, and many others. The key problem in solution of various applied problems is that of the stability of solutions of systems of equations in various senses. The methods of the classical stability theory, if appropriately adapted, may be applied to systems containing a small parameter.

The progress in solving problems of the theory of stability and nonlinear perturbations is associated with finding ways, around significant difficulties connected with the growth of the number of variables characterizing the state of a system, which may include critical variables. In addition, the presence of critical variables may result in a situation when not only the first approximation cannot solve a stability problem, but also the further nonlinear approximations below some order cannot solve it.

New approaches recently developed for systems with a small parameter include the following:

A. The development of the direct Lyapunov method for the study of the boundedness and stability of systems with a finite number of degrees of freedom with respect to two different measures.

B. The analysis of stability on the basis of the combination of the concepts of the direct Lyapunov method and the averaging method of nonlinear mechanics for some classes of linear and nonlinear systems.

C. The generalization of the direct Lyapunov method on the basis of the concepts of the comparison principle and the averaging method of nonlinear mechanics.

D. The development of the method of matrix-valued Lyapunov functions and its application in the study of stability of singularly perturbed systems.

The core subject of investigation in this book is the systems with a small parameter, including nonlinear systems of weakly connected equations. Here approaches A – D are applied and developed when solving specifically defined problems.

The monograph consists of five chapters, and their content is outlined below.

The first chapter provides description of the mathematical foundations of the methods of qualitative analysis of systems with small parameter. Namely, it contains the necessary information from the theory of integral and differential inequalities, the comparison technique, and the main theorems of the direct Lyapunov method. In this chapter, stability definitions for systems with small parameter are discussed as well as their relationship with the classical Lyapunov definitions.

The second chapter contains the results of the development of new approaches to the problem of the boundedness of motion of weakly connected nonlinear systems. The direct Lyapunov method and the comparison technique are applied in this chapter to establish the conditions for the boundedness of nonlinear systems with respect to two different measures. The results of the analysis of the dynamical behavior of an individual subsystem in a complex system of weakly connected equations are given, which were obtained via the application of strengthened Lyapunov functions. The uniform boundedness and the uniform ultimate boundedness are in terms of the vector Lyapunov function and the theory of M-matrices. The final section of the chapter deals with the problem of the boundedness of solutions of the Lienard oscillator with weak damping, the boundedness of solutions of the Lurie–Postnikov system, and the boundedness of solutions for nonlinear systems with weak nonlinear constraints.

In the third chapter, the application of the direct Lyapunov method and the comparison technique is set forth for solving the problem of stability of solutions of a weakly connected system of differential equations. The analysis is carried out under different assumptions on the connection functions of subsystems. The sufficient conditions for asymptotic and uniform asymptotic stability were established based on the auxiliary vector function. Also, a general problem on polystability of motion of a nonlinear system with small parameter is formulated. In the section dealing with applications, some problems of the automatic control theory are considered.

The fourth chapter contains the description of one general approach to the study of stability of solutions for nonlinear systems with small perturbing forces. This approach is based on the generalization of the direct Lyapunov method combined with the asymptotic method of nonlinear mechanics. In addition, generalizations of the main Lyapunov theorems on stability and Chetayevs theorem on instability for the class of systems under consideration are given. Due to the difficulties arising in construction of solutions for a degenerate system, an approach associated with the substitution for its exact solution by the solution of a limiting system is discussed in the chapter. Systems of weakly connected oscillators are considered as applications.

The fifth chapter is dedicated to the analysis of stability of singularly perturbed systems. For this purpose the generalization of the direct Lyapunov method on the basis of matrix-valued functions is applied. The proposed approach enables one to reduce the requirements to the dynamic properties of a singular system and the boundary layer and obtain new conditions for the asymptotic stability of the initial system. The proposed approach is applied in the analysis of stability of large-scale singularly perturbed systems. As an engineering application, the singularly perturbed Lurie–Postnikov system is considered.

The fifth chapter gives an account of results of the analysis of hybrid systems with weakly connected subsystems on the basis of the generalization of the direct Lyapunov method. Both vector and matrix-valued auxiliary functions are applied here.

Thus, this book contains the description of the main approaches to the analysis of stability of solutions of systems with a small parameter, which have been developed for the last four decades. Those approaches do not exhaust the given problem, but they may help solve many applied problems of the modern technique and technologies. In addition, their certain "incompleteness" leaves room for further search of more effective approaches in this line of investigation.

The Member of the Academy of Sciences Yu. A. Mitropolsky has more than once drawn the authors' attention to the necessity of the development of methods of qualitative analysis of systems with a small parameter that could be used in engineering. The proposed approaches provide a kind of answer to this problem in the context of the modern development of the qualitative theory of equations.

The main results given in this monograph were obtained in the department of stability of processes of the Institute of Mechanics of the National Academy of Sciences of Ukraine in the period of 1978–2010 in the frame of the department's subject of scientific investigation: *the development of qualitative and analytical methods of the analysis of dynamics and stability of functioning of complex nonlinear and controllable systems, including systems with structural and stochastic perturbations, and with aftereffect.*

This subject is the extension and further development of some lines of investigation conducted in the institute by Member of the Academy of Sciences N.M. Krylov and N.N. Bogolyubov[1].

In the process of the above investigation, some issues were discussed with Professors T.A. Barton, V.I. Zubov, V. Lakshmikantham, V.M. Starzhinsky, D.D. Šiljak, and others.

Separate parts of this book were written by: A.A. Martynyuk (Chapters 1–3, 5); L.N. Chernetskaya (Chapters 1, 4), and V.A. Martynyuk (Chapters 2, 3). The chapters attributed to two authors have been written by both of them together.

---

[1] See the bibliography of the works of N. N. Bogolyubov in nonlinear mechanics in the article of A. A. Martynyuk, E. F. Mishchenko, A. M. Samoilenko, and A. D. Sukhanov, Academician N.N. Bogoliubov, *Nonlinear Dynamics and Systems Theory*, **9**(2) (2009) 109–115.

Acknowledgments

We thank all of our teachers and collaborators who were involved in our work over the book. We have published some parts of this book in the context of a research project at the S.P. Timoshenko Institute of Mechanics NAS of Ukraine, and we thank the Institute for the support we had in the process of our work.

We wish, first of all, to express considerable gratitude to academician Yu. A. Mitropolskiy who has given us considerable encouragement to begin studies and research of weakly connected nonlinear systems. Professors T.A. Burton, N.Z. Litvin-Sedoi, G.I. Melnikov, V.M. Starzhinsky, and V.I. Zubov must be thanked for discussion and comments of some parts of the book and our journal publications in the direction.

Finally, we wish to acknowledge the cooperation of Bob Stern (CRC Press, Taylor & Francis Group) and we appreciate the consideration given to this book by the editors of the series, Pure and Applied Mathematics.

Chapter 1

Preliminaries

1.1 Introductory Remarks

For the mathematical simulation of many processes of the real world, systems of nonlinear differential equations are used which contain a small positive parameter μ participating in the determination of the time scale of the process. In general, the equations have the form

$$\frac{dy}{dt} = Y(t, y, \mu), \quad y(t_0) = y_0, \tag{1.1.1}$$

where $y(t) \in R^n$ is the vector of the state of a system at a point of time $t \in R$, $R = (-\infty, +\infty)$, $t_0 \in R_+$, $Y \colon R \times R^n \times M \to R^n$, $M = (0, 1]$, μ is a small parameter.

Section 1.2 contains Gronwall-Bihari fundamental inequalities, differential and integral inequalities applied in the book.

In Section 1.3, theorems on the dependence of solutions of a system of differential equations on the parameter μ are formulated. Here it is also demonstrated that the application of a finite majorizing equation instead of a differential equation essentially improves the estimate of the radius of convergence of series, which represent a solution of a system in powers of the parameter. In addition, the well-known theorems of extendability of solutions are quoted, as well as the Poincare theorem on presentation of solutions in powers of a small parameter.

In Section 1.4, the original Lyapunov definitions of various types of stability of motion are given.

Section 1.5 contains the basic theorems of Lyapunov direct method on the basis of a scalar Lyapunov function.

Section 1.6 contains the basic theorems of the principle of comparison with the scalar and vector Lyapunov functions.

In the concluding section, the definitions of the main types of μ-stability of solutions of a system with a small parameter are given. The connection between Lyapunov stability and μ-stability is demonstrated with a number of specific examples.

Thus, Chapter 1 contains a set of known and new results which constitute the basis for the mathematical analysis of solutions of systems of differential equations with a small parameter.

1

1.2   Fundamental Inequalities

1.2.1   Gronwall type inequalities

To begin with, we take one of the simplest integral inequalities that are used most often.

Theorem 1.2.1 *Let the functions $m, v \in C(R_+, R_+)$. Assume that for some $c \geq 0$ the following inequality holds:*

$$m(t) \leq c + \int_{t_0}^{t} v(s)m(s)\,ds, \quad t \geq t_0 \geq 0. \tag{1.2.1}$$

Then

$$m(t) \leq c \exp\left[\int_{t_0}^{t} v(s)\,ds\right], \quad t \geq t_0. \tag{1.2.2}$$

Proof Denote the right-hand part of the inequality (1.2.1) by $z(t)$. Here $z(t_0) = c$, $m(t) \leq z(t)$, and $z'(t) = v(t)m(t) \leq v(t)z(t)$ at all $t \geq t_0$. Since

$$z'(t) \exp\left(-\int_{t_0}^{t} v(s)\,ds\right) - z(t)v(t)\exp\left(-\int_{t_0}^{t} v(s)\,ds\right)$$

$$= \frac{d}{dt}\left(z(t)\exp\left(-\int_{t_0}^{t} v(s)\,ds\right)\right),$$

then

$$\frac{d}{dt}\left(z(t)\exp\left(-\int_{t_0}^{t} v(s)\,ds\right)\right) \leq 0.$$

Integrating this inequality from t_0 to t, obtain

$$z(t)\exp\left(-\int_{t_0}^{t} v(s)\,ds\right) - z(t_0) \leq 0.$$

Taking into account that $z(t_0) = c$ and $m(t) \leq z(t)$, we obtain the inequality (1.2.2) at any $c \geq 0$. Theorem 1.2.1 is proved.

The above classical proof of Theorem 1.2.1 holds much significance. However, we can prove this theorem by using a linear differential inequality and a formula of variation of constants. For this purpose, consider a more general case.

Theorem 1.2.2 *Let the functions* $m, v, h \in C(R_+, R_+)$ *and*

$$m(t) \le h(t) + \int_{t_0}^{t} v(s)m(s)\,ds, \quad t \ge t_0. \tag{1.2.3}$$

Then

$$m(t) \le h(t) + \int_{t_0}^{t} [v(s)h(s)] \exp\left(\int_{s}^{t} v(\xi)\,d\xi\right) ds, \quad t \ge t_0. \tag{1.2.4}$$

If the function h *is differentiable, then*

$$m(t) \le h(t_0) \exp\left(\int_{t_0}^{t} v(s)\,ds\right) + \int_{t_0}^{t} h'(s) \exp\left(\int_{s}^{t} v(\xi)\,d\xi\right) ds, \quad t \ge t_0. \tag{1.2.5}$$

Proof To prove the inequality (1.2.4), assume that $p(t) = \int_{t_0}^{t} v(s)m(s)\,ds$, so that $p(t_0) = 0$ and

$$p'(t) = v(t)m(t), \quad t \ge t_0.$$

Since $m(t) \le h(t) + p(t)$, then

$$p'(t) \le v(t)p(t) + v(t)h(t), \quad t \ge t_0.$$

Assuming $q(t) = p(t) \exp\left(-\int_{t_0}^{t} v(s)\,ds\right)$, we see that $q(t_0) = 0$ and

$$q'(t) = [p'(t) - v(t)p(t)] \exp\left(-\int_{t_0}^{t} v(s)\,ds\right) \le h(t)v(t) \exp\left(-\int_{t_0}^{t} v(s)\,ds\right),$$

whence

$$q(t) \le \int_{t_0}^{t} h(s)v(s) \exp\left(-\int_{t_0}^{t} v(\xi)\,d\xi\right) ds, \quad t \ge t_0.$$

As a result, obtain the inequality

$$p(t) \le \int_{t_0}^{t} v(s)h(s) \exp\left(\int_{s}^{t} v(\xi)\,d\xi\right) ds, \quad t \ge t_0,$$

which directly implies (1.2.4).

To prove the inequality (1.2.5), denote the right-hand part of the inequality (1.2.3) by $p(t)$, so that

$$p'(t) = v(t)m(t) + h'(t), \quad p(t_0) = h(t_0).$$

The above expression, in view of (1.2.3), results in the linear differential inequality

$$p'(t) \leq v(t)p(t) + h'(t), \quad p(t_0) = h(t_0).$$

Now it is easy to obtain

$$p(t) \leq h(t_0) \exp \left(\int_{t_0}^{t} v(s)\, ds \right) + \int_{t_0}^{t} h'(s) \exp \left(\int_{s}^{t} v(\xi)\, d\xi \right) ds, \quad t \geq t_0,$$

hence the estimate (1.2.5).

At first, the estimates (1.2.4) and (1.2.5) are different. In actual truth, they are equivalent. Thus, integrating the second summand in the right-hand part of the inequality (1.2.5) in parts, obtain

$$\int_{t_0}^{t} h'(s) \exp \left(\int_{s}^{t} v(\xi)\, d\xi \right) ds = h(t) - h(t_0) \exp \left(\int_{t_0}^{t} v(\xi)\, d\xi \right) +$$

$$+ \int_{t_0}^{t} h(s)v(s) \exp \left(\int_{s}^{t} v(\xi)\, d\xi \right) ds,$$

which, taking into account the estimate (1.2.5), gives (1.2.4). Thus, it is obvious that the assumption of the differentiability of the function $h(t)$ does not give anything new.

If we assume in Theorem 1.2.2 that the function h is positive and nondecrescent, then the estimate (1.2.4) can be transformed into

$$m(t) \leq h(t) \exp \left(\int_{t_0}^{t} v(s)\, ds \right), \quad t \geq t_0. \tag{1.2.6}$$

Assuming $w(t) = \dfrac{m(t)}{h(t)}$, from (1.2.3) obtain

$$w(t) \leq 1 + \int_{t_0}^{t} v(s)w(s)\, ds, \quad t \geq t_0.$$

Hence, according to Theorem 1.2.1, obtain $w(t) \leq \exp \left(\int_{t_0}^{t} v(s)\, ds \right)$, which implies (1.2.6).

The estimate (1.2.6) can be obtained from the inequality (1.2.4), since

$$h(t) + \int_{t_0}^{t} v(s)h(s) \exp\left(\int_{t_0}^{t} v(\xi)\,d\xi\right) ds$$

$$\leq h(t)\left[1 + \int_{t_0}^{t} v(s) \exp\left(\int_{s}^{t} v(\xi)\,d\xi\right) ds\right]$$

$$= h(t)\left(1 - \int_{s}^{t} e^{\sigma(s)}\,d\sigma(s)\right) = h(t)\exp\left(\int_{t_0}^{t} v(\xi)\,d\xi\right), \quad t \geq t_0.$$

Example 1.2.1 Let the function $m \in C(R_+, R_+)$ and

$$m(t) \leq \int_{t_0}^{t} (a + bm(s))\,ds, \quad t \geq t_0,$$

where $a \geq 0$ and $b > 0$. Then

$$m(t) \leq \frac{a}{b}\left[\exp(b(t - t_0)) - 1\right], \quad t \geq t_0.$$

Example 1.2.2 Let the function $m \in C(R_+, R_+)$ and

$$m(t) \leq a + \int_{t_0}^{t} (b + cm(s))\,ds, \quad t \geq t_0,$$

where $a, b \geq 0$ and $c > 0$. Then

$$m(t) \leq \frac{b}{c}\left[\exp(c(t - t_0)) - 1\right] + a\exp(c(t - t_0)), \quad t \geq t_0.$$

Example 1.2.3 Let the function $m \in C(R_+, R_+)$ and

$$m(t) \leq e^t + b\int_{t_0}^{t} m(s)\,ds, \quad t \geq t_0$$

at $b > 0$. Then

$$m(t) \leq \exp[(b + 1)t - bt_0], \quad t \geq t_0.$$

Example 1.2.4 Let the function $m \in C(R_+, R_+)$ and let for $t \geq t_0$ the following inequality hold:

$$m(t) \leq m(t_0)\exp(-r(t - t_0)) + \int_{t_0}^{t} [\exp(-r(t - s))](am(s) + b)\,ds,$$

where $r, a, b \geq 0$ and $r - a > 0$. Then for $t \geq t_0$

$$m(t) \leq m(t_0) \exp[-(r-a)(t-t_0)] + \frac{b}{r-a}[1 - \exp(-(r-a)(t-t_0))].$$

Now we will consider integral inequalities with a separable kernel, as they may also be reduced to linear differential inequalities.

Theorem 1.2.3 *Let the functions* $m, h, q, v \in C(R_+, R_+)$ *and the following inequality be satisfied:*

$$m(t) \leq h(t) + \int_{t_0}^{t} q(t)v(s)m(s)\, ds, \quad t \geq t_0. \tag{1.2.7}$$

Then

$$m(t) \leq h(t) + q(t) \int_{t_0}^{t} v(s)h(s) \exp\left(\int_{s}^{t} v(\xi)\, d\xi\right) ds, \quad t \geq t_0. \tag{1.2.8}$$

Proof Assume that $p(t) = \int_{t_0}^{t} v(s)m(s)\, ds$, so that $p(t_0) = 0$ and $p'(t) = v(t)m(t)$. Since $m(t) \leq h(t) + q(t)p(t)$, obtain

$$p'(t) \leq v(t)q(t)p(t) + v(t)h(t), \quad t \geq t_0,$$

hence

$$p(t) \leq \int_{t_0}^{t} v(s)h(s) \exp\left(\int_{s}^{t} v(\xi)q(\xi)\, d\xi\right) ds, \quad t \geq t_0.$$

Hence follows the inequality (1.2.8).

Corollary 1.2.1 Let the functions $m, h, g_i, v_i \in C(R_+, R_+)$, $i = 1, 2, \ldots, n$, and

$$m(t) \leq h(t) + \sum_{i=1}^{n} g_i(t) \int_{t_0}^{t} v_i(s)m(s)\, ds, \quad t \geq t_0.$$

Then

$$m(t) \leq h(t) + G(t) \int_{t_0}^{t} V(s)h(s) \exp\left(\int_{s}^{t} V(\xi)G(\xi)\, d\xi\right) ds, \quad t \geq t_0,$$

where $G(t) = \sup_i g_i(t)$ and $V(t) = \sum_{i=1}^{n} v_i(t)$.

1.2.2 Bihari type inequalities

The theory of Gronwall integral inequalities considered in Subsection 1.2.1 may be applied to a separate type of nonlinear integral inequality known as Bihari inequalities. In this section you will find some results related to such inequalities, which correspond to the results obtained in Subsection 1.2.1.

Theorem 1.2.4 *Let the functions* $m, v \in C(R_+, R_+)$, $g \in C((0, \infty), (0, \infty))$ *and let* $g(u)$ *be nondecrescent with respect to* u. *Assume that for some* $c > 0$

$$m(t) \leq c + \int_{t_0}^{t} v(s)g(m(s))\, ds, \quad t \geq t_0 > 0. \tag{1.2.9}$$

Then the following inequality holds:

$$m(t) \leq G^{-1}\left[G(c) + \int_{t_0}^{t} v(s)\, ds\right], \quad t_0 \leq t < T,$$

where $G(u) - G(u_0) = \int_{u_0}^{u} \frac{ds}{g(s)}$, $G^{-1}(u)$ *is the reverse function to* $G(u)$ *and* $T = \sup\left\{t \geq t_0 : G(c) + \int_{t_0}^{t} v(s)ds \in \operatorname{dom} G^{-1}\right\}$.

Proof Denote the right-hand part of the inequality (1.2.9) by $p(t)$ so that $p(t_0) = c$ and $p'(t) = v(t)g(m(t))$. Since g is nondecrescent with respect to u and $m(t) \leq p(t)$, then $p'(t) \leq v(t)g(p(t))$, $p(t_0) = c$. Integrating this inequality from t_0 to t, obtain

$$G(p(t)) - G(c) = \int_{c}^{p(t)} \frac{dz}{g(z)} \leq \int_{t_0}^{t} v(s)\, ds$$

and therefore

$$m(t) \leq p(t) \leq G^{-1}\left[G(c) + \int_{t_0}^{t} v(s)\, ds\right], \quad t_0 \leq t < T.$$

The function $g \in C[R_+, R_+]$ is said to be subadditive if $g(u + v) \leq g(u) + g(v)$, and superadditive if the above inequality has the opposite sign.

Theorem 1.2.5 *Let the functions* $m, v, h \in C(R_+, R_+)$, $g \in C((0, \infty), (0, \infty))$, *let the function* $g(u)$ *be nondecrescent, and let the following inequality hold:*

$$m(t) \leq h(t) + \int_{t_0}^{t} v(s)g(m(s))\, ds, \quad t \geq t_0.$$

Then:

(a) *if the function $g(u)$ is subadditive, then*

$$m(t) \le h(t) + G^{-1}\left[G(c) + \int\limits_{t_0}^{t} v(s)\,ds\right], \quad t_0 \le t \le T_0 < T, \quad (1.2.10)$$

where G, G^{-1}, and T have the same values as in Theorem 1.2.4, $c = \int\limits_{t_0}^{T_0} v(s)g(h(s))\,ds$;

(b) *if the function h is not increscent, then*

$$m(t) \le -h(t_0) + G^{-1}\left\{G[h(t_0)] + \int\limits_{t_0}^{t} v(s)\,ds\right\}, \quad t_0 \le t < T. \quad (1.2.11)$$

Proof Assuming that $p(t) = \int\limits_{t_0}^{t} v(s)g(m(s))\,ds$ and taking into account the properties of g, obtain $p(t_0) = 0$ and

$$p'(t) \le v(t)g(p(t)) + v(t)g(h(t)).$$

Note that the function $\sigma(t) = \int\limits_{t_0}^{t} v(s)g(h(s))\,ds$ is nondecrescent, and hence, assuming $c = \sigma(T_0)$, at some T_0, $\quad t_0 \le T_0 < T$, obtain

$$p(t) \le c + \int\limits_{t_0}^{t} v(s)g(p(s))ds, \quad t_0 \le t \le T_0 < T.$$

According to Theorem 1.2.4, the above expression implies the estimate (1.2.10).

If the function h is not increscent, then the definition of $p(t)$ implies that $g(m(t)) \le g(h(t_0) + p(t))$. Assuming $h(t_0) + p(t) = w(t)$, obtain

$$w(t) = p'(t) = v(t)g(m(t)) \le v(t)g(w(t)), \quad w(t_0) = h(t_0).$$

From the above equation, according to Theorem 1.2.9, we arrive at the estimate (1.2.11).

The estimate (1.2.10) may also be obtained in the case when in Theorem 1.2.5 the function $g(u)$ is assumed to be nonincrescent and superadditive with respect to u.

Using Theorem 1.2.2, the following result may be proved.

Theorem 1.2.6 *Let the functions* $m, h \in C(R_+, R_+)$, $g \in C((0, \infty), (0, \infty))$ *and the function* $g(u)$ *be nondecrescent with respect to* u. *Assume that* $K \in C[R_+^2, R_+]$, *there exists a function* $K_t(t, s)$ *which is continuous and nonnegative, and at* $t \geq t_0$ *the following inequality holds:*

$$m(t) \leq h(t) + \int_{t_0}^{t} K(t, s)g(m(s))\, ds.$$

Then:

(a) *if the function* g *is subadditive, then*

$$m(t) \leq h(t) + v_2(t) + G^{-1}\left[G(c) + \int_{t_0}^{t} v_1(s)\, ds\right], \quad t_0 \leq t \leq t_0 < T,$$

where G, G^{-1} *and* T *have the same values as in Theorem 1.2.4,*

$$c = \int_{t_0}^{T_0} v_1(s)g(v_2(s))\, ds, \quad v_1(t) = K(t, t) + \int_{t_0}^{t} K_t(t, s)\, ds,$$

$$v_2(t) = K(t, t)h(t) + \int_{t_0}^{t} K_t(t, s)g(h(s))\, ds;$$

(b) *if the function* h *is nondecrescent, then*

$$m(t) \leq h(t) - h(t_0) + G^{-1}\left[G(h(t_0)) + \int_{t_0}^{t} v_1(s)\, ds\right], \quad t_0 \leq t < T.$$

A typical nonlinear integral inequality that can be reduced to Theorem 1.2.4. has the following form.

Theorem 1.2.7 *Let the functions* $m, v \in C(R_+, R_+)$, $w \in C(R_+^2, R_+)$ *and let the following inequalities hold:*

$$m(t) \leq c + \int_{t_0}^{t} \{v(s)m(s) + w[s, m(s)]\}\, ds, \quad t \geq t_0, \tag{1.2.12}$$

where $c > 0$. *Assume that*

$$w\left(t, z\exp\left(\int_{t_0}^{t} v(s)\, ds\right)\right) \leq \lambda(t)g(z)\exp\left(\int_{t_0}^{t} v(s)\, ds\right), \tag{1.2.13}$$

where $\lambda \in C(R_+, R_+)$, $g \in C((0, \infty), (0, \infty))$ *and the function* $g(u)$ *is decrescent with respect to* u.

Then

$$m(t) \leq G^{-1}\left[G(c) + \int_{t_0}^{t} \lambda(s)\, ds\right] \exp\left(\int_{t_0}^{t} v(s)\, ds\right), \quad t_0 \leq t < T, \quad (1.2.14)$$

G, G^{-1}, *and* T *are the same as in Theorem 1.2.4.*

Proof Let the right-hand part of the inequality (1.2.12) be equal to $p(t) \exp\left(\int_{t_0}^{t} v(s)\, ds\right)$, so that, using (1.2.12) and (1.2.13), obtain

$$[p'(t) + v(t)p(t)] \exp\left(\int_{t_0}^{t} v(s)\, ds\right) = v(t)m(t) + w(t, m(t))$$

$$\leq [v(t)p(t) + \lambda(t)g(m(t))] \exp\left(-\int_{t_0}^{t} v(s)\, ds\right) \exp\left(\int_{t_0}^{t} v(s)\, ds\right).$$

Since the function g is nondecrescent and $m(t) \leq p(t) \exp\left(\int_{t_0}^{t} v(s)\, ds\right)$, obtain

$$p'(t) \leq \lambda(t)g(p(t)), \quad p(t_0) = c.$$

Hence according to Theorem 1.2.4 obtain

$$p(t) \leq G^{-1}\left[G(c) + \int_{t_0}^{t} \lambda(s)\, ds\right], \quad t_0 \leq t < T,$$

which proves the estimate (1.2.14).

Example 1.2.5 Let the functions $m, v, h \in C(R_+, R_+)$, so that at $c > 0$, $0 \leq p < 1$

$$m(t) \leq c + \int_{t_0}^{t} v(s)m(s)\, ds + \int_{t_0}^{t} h(s)(m(s))^p ds, \quad t \geq t_0.$$

Then at $t \geq t_0$

$$m(t) \leq \left\{c^q + q \int_{t_0}^{t} h(s) \exp\left[q \int_{t_0}^{s} v(\xi)\, d\xi\right] ds\right\}^{1/q} \exp\left[\int_{t_0}^{t} v(s)\, ds\right],$$

where $q = 1 - p$.

If the kernel $K(t, s)$ in Theorem 1.2.6 is such that $K_t(t, s) \leq 0$, then in the frame of this theorem one only can obtain a rough estimate. However, the following theorem provides the possibility to obtain a better estimate.

Theorem 1.2.8 *Assume that* $m \in C(R_+, R_+)$, $g \in C((0, \infty), (0, \infty))$, *the function* $g(u)$ *is nondecrescent with respect to* u, *and at some* $c > 0$, $\alpha > 0$

$$m(t) \leq c + \int_{t_0}^{t} e^{-\alpha(t-s)} g(m(s)) \, ds, \quad t \geq t_0. \tag{1.2.15}$$

Then

$$m(t) \leq (1 + \lambda_0)c, \quad t \geq t_0,$$

where $\lambda_0 > 0$ *satisfies the relations*

$$g((1 + \lambda_0)c) - \alpha c \lambda_0 = 0, \quad g((1 + \lambda)c) - \alpha c \lambda > 0, \quad \lambda \in [0, \lambda_0). \tag{1.2.16}$$

Proof Let the right-hand part of the inequality (1.2.15) be equal to $p(t)$, so that $p(t_0) = c$ and

$$p'(t) = g(m(t)) - \alpha(p(t) - c) \leq g(p(t)) - \alpha p(t) + \alpha c. \tag{1.2.17}$$

Transforming $p(t) = (1 + z(t))c$ and $\tau = \alpha t$, it is easy to reduce (1.2.17) to the form

$$\frac{dz}{d\tau} \leq \frac{1}{\alpha c} g((1 + z)c) - z, \quad z\left(\frac{\tau_0}{\alpha}\right) = 0.$$

We state that $z\left(\frac{\tau}{\alpha}\right) \leq \lambda_0$, $\tau \in [\tau_0, \infty)$. If this is not so, then one can find a $\tau^* < \infty$ such that $z\left(\frac{\tau^*}{\alpha}\right) = \lambda_0$ and $z\left(\frac{\tau}{\alpha}\right) \leq \lambda_0$, $\tau \in [\tau_0, \tau^*]$. From (1.2.17) it is clear that $\frac{1}{\alpha c} g((1 + \lambda)c) < \lambda_0$ for all $\lambda \in [0, \lambda_0]$. Hence

$$\infty = \int_0^{\lambda_0} \frac{ds}{\lambda_0 - s} \leq \int_0^{z(\frac{\tau^*}{\alpha})} \frac{ds}{\frac{1}{\alpha c} g((1 + s)c) - s} \leq \tau^* - \tau_0,$$

which is a contradiction. Therefore, $z\left(\frac{\tau}{\alpha}\right) \leq \lambda_0$ at all $\tau \in [\tau_0, \infty)$, which in its turn results in the inequality

$$m(t) \leq p(t) \leq (1 + \lambda_0)c, \quad t \geq t_0.$$

1.2.3   Differential inequalities

It is known that the theory of differential equations plays a key role in the study of qualitative behavior of solutions of differential equations of different types. This theory is helpful when one uses integral and integrodifferential equations, since in numerous cases their study may be reduced to the study of differential inequalities.

Consider the differential system

$$\frac{du}{dt} = g(t, u), \quad u(t_0) = u_0, \tag{1.2.18}$$

where $g \in C(R_+ \times R^n, R^n)$.

The function g is called quasimonotone nondecrescent, if from the component-wise inequality $x \leq y$ and $x_i = y_i$ at some i, $1 \leq i \leq n$, it follows that $g_i(t, x) \leq g_i(t, y)$ at any $t \geq t_0$.

Now we will consider the concept of an extremum solution of the system (1.2.18).

Let $r(t)$ be a solution of the system (1.2.18), existing on some interval $J = [t_0, t_0 + a]$. Then $r(t)$ is called the maximum solution of the system (1.2.18), if for each solution $u(t)$ of the system (1.2.18), existing on J, the following inequality holds:

$$u(t) \leq r(t), \quad t \in J. \tag{1.2.19}$$

The minimum solution is determined in a similar way, but the sign in the inequality (1.2.19) is changed to the opposite sign.

Surely, inequalities between vectors are understood componentwise.

For our study, the following known result is required the proof of which can be found in Walter [1].

Theorem 1.2.9   *Let the function* $g \in C(E, R^n)$, *where E is an open set (t, u) in R^{n+1} and $g(t, u)$ is a quasimonotone nondecrescent function with respect to u for each t.*

Then:

(a) *if $(t_0, u_0) \in E$, then the system (1.2.18) has an extremum solution which can be extended to the boundary E;*

(b) *if J is an interval of existence of the maximum solution $r(t)$ of the system (1.2.18) on any compact interval $[t_0, T]$, then there exists $\varepsilon_0 > 0$ such that at any $0 < \varepsilon < \varepsilon_0$ solutions $u(t, \varepsilon)$ of the system*

$$\frac{du}{dt} = g(t, u) + \varepsilon, \quad u(t_0) = u_0 + \varepsilon \tag{1.2.20}$$

exist on the interval $[t_0, T]$ and $\lim_{\varepsilon \to 0} u(t, \varepsilon) = r(t)$ uniformly on $[t_0, T]$.

Lemma 1.2.1 *Let the functions* $v, w \in C(J, R)$ *and for some fixed Dini derivative* $Dv(t) \le w(t)$, $t \in J \setminus S$, *where* S *is a countable subset of* J. *Then* $D_- v(t) \le w(t)$ *on* J.

The following result of comparison in the scalar form contains the key concept of the inequality theory.

Theorem 1.2.10 *Let the function* $g \in C(R_+ \times R_+, R)$ *and let* $r(t)$ *be the maximum solution of the system (1.2.18), which exists on the interval* $[t_0, \infty)$. *Assume that* $m \in C(R_+, R_+)$ *and* $Dm(t) \le g(t, m(t))$, $t \ge t_0$, *where* D *is any fixed Dini derivative.*
Then the inequality $m(t_0) \le u_0$ *implies that* $m(t) \le r(t)$ *at all* $t \ge t_0$.

Proof According to Lemma 1.2.1, obtain

$$D_- m(t) \le g(t, m(t)), \quad t \ge t_0,$$

where $D_- m(t) = \liminf\{[m(t + \theta) - m(t)]\theta^{-1} \colon \theta \to 0^-\}$. Let $t_0 < T < \infty$. According to Theorem 1.2.9, the solution $u(t, \varepsilon)$ of the system (1.2.20) exists on the interval $[t_0, T]$ at all sufficiently small $\varepsilon > 0$ and $\lim_{\varepsilon \to 0} u(t, \varepsilon) = r(t)$ uniformly on $[t_0, T]$. Hence it suffices to show that

$$m(t) < u(t, \varepsilon), \quad t \in [t_0, T]. \tag{1.2.21}$$

If the inequality (1.2.21) does not hold, then there exists such a value $t_1 \in [t_0, T]$, that

$$m(t_1) = u(t_1, \varepsilon), \quad m(t) \ge u(t, \varepsilon), \quad t \in [t_0, t_1].$$

Hence obtain

$$D_- m(t_1) \ge u'(t_1, \varepsilon),$$

which, in its turn, results in the contradiction

$$g(t_1, m(t_1)) \ge D_- m(t_1) \ge u'(t_1, \varepsilon) = g(t_1, u(t_1, \varepsilon)) + \varepsilon.$$

Therefore the inequality (1.2.21) holds, which completes the proof of the theorem.

To avoid the repetition of the proof, we have not considered the lower estimate for $m(t)$, which can be obtained by the change of signs in the inequalities for opposite signs. To continue the discussion we will need the following theorem which contains the lower estimate for $m(t)$.

Theorem 1.2.11 *Let the function* $g \in C(R_+ \times R_+, R)$ *and let* $\rho(t)$ *be the minimum solution of the system (1.2.18) existing on* $[t_0, \infty)$. *Assume that* $m \in C(R_+, R_+)$ *and* $Dm(t) \ge g(t, m(t))$, $t \ge t_0$, *where* D *is any fixed Dini derivative.*
Then the inequality $m(t_0) \ge u_0$ *implies that* $m(t) \ge \rho(t)$ *at all* $t \ge t_0$.

The proof of the above theorem is similar to that of Theorem 1.2.10. Instead of solutions of the system (1.2.20), here we consider the solutions $v(t, \varepsilon)$ of the system

$$v' = g(t, v) - \varepsilon, \quad v(t_0) = u_0 - \varepsilon$$

for a sufficiently small $\varepsilon > 0$ on an interval $[t_0, T]$, and $\lim_{\varepsilon \to 0} v(t, \varepsilon) = \rho(t)$ uniformly on $[t_0, T]$. To complete the proof, it is sufficient to see that

$$m(t) > v(t, \varepsilon), \quad t \in [t_0, T].$$

When using Theorem 1.2.10, it is necessary that the function g should be quasimonotone nondecrescent, which is a necessary condition for the existence of extremum solutions of the system (1.2.18). Thus, we obtain the following extension of Theorem 1.2.10.

Theorem 1.2.12 *Let $g \in C(R_+ \times R_+^n, R^n)$, $g(t, u)$ be a quasimonotone function, nondecrescent with respect to u for each t, and let $r(t)$ be the maximum solution of the system (1.2.18), existing on $[t_0, \infty)$. Assume that the inequality $Dm(t) \leq g(t, m(t))$, $t \geq t_0$, holds for a fixed Dini derivative.*

Then, from the inequality $m(t_0) \leq u_0$, it follows that $m(t) \leq r(t)$ at all $t \geq t_0$.

As noted above, the inequalities in Theorem 1.2.12 are componentwise.

Instead of considering those inequalities between vectors, we will use the concept of a cone in order to introduce partial ordering on R^n and prove Theorem 1.2.12 in such a frame. Obviously, such an approach is more general and used for cone-valued functions. Therefore, the extension of the theory of differential inequalities is a result corresponding to Theorem 1.2.12 in arbitrary cones.

The subset $K \subset R^n$ is called a cone if it has the following properties:

$$\lambda K \subset K, \quad \lambda \geq 0, \quad K + K \subset K,$$
$$K = \overline{K}, \quad K \cap \{-K\} = \{0\}, \quad K^0 \neq \varnothing, \tag{1.2.22}$$

where $\overline{K}$ is the closure of K, K^0 is the interior of the cone K.

Let ∂K denote the boundary of the cone K. By the cone K the ordering relationship in R^n is introduced, which is determined by the relations

$$x \leq y \quad \text{if and only if} \quad y - x \in K,$$
$$x < y \quad \text{if and only if} \quad y - x \in K^0. \tag{1.2.23}$$

The set K^*, defined as $K^* = \{\varphi \in R^n : \varphi(x) \geq 0 \text{ at all } x \in K\}$, where the function $\varphi(x)$ denotes the scalar product $\langle \varphi, x \rangle$ and is called an adjoint cone, satisfies the conditions (1.2.22).

Note that $K = (K^*)^*$, $x \in K^0$, if and only if $\varphi(x) > 0$ at all $\varphi \in K_0^*$, and $x \in \partial K$, if and only if $\varphi(x) = 0$ at some $\varphi \in K_0^*$, where $K_0 = K \setminus \{0\}$.

Now we can give the definition of the property of quasimonotonicity with respect to the cone K.

The function $f \in C[R^n, R^n]$ is quasimonotone nondecrescent with respect to the cone K, if from $x \le y$ and $\varphi(x-y)=0$ at some $\varphi \in K_0^*$ it follows that $\varphi(f(x)-f(y)) \le 0$.

If the function f is linear, that is, $f(x) = Ax$, where A is an $(n \times n)$-matrix, then the property of quasimonotonicity of the function f means that the conditions $x \ge 0$ and $\varphi(x)=0$ at some $\varphi \in K_0^*$ follow from $\varphi(Ax) \ge 0$.

If $K = R_+^n$, then the quasimonotonicity of f amounts to the above definition.

For an ordinary cone Theorem 1.2.9 is true. Note that it is possible to prove the existence of extremum solutions for differential equations in a Banach space as well. Now we will prove the result of comparison with respect to the cone K.

Theorem 1.2.13 *Let the vector function $g \in C(R_+ \times R^n, R^n)$ and $g(t,u)$ be a quasimonotone function, nondecrescent with respect to u relative to the cone K for each $t \in R_+$. Let $r(t)$ be the maximum solution of the system (1.2.18) with respect to the cone K existing on the interval $[t_0, \infty)$, and for $t \ge t_0$*

$$D_-m(t) \le g(t, m(t)), \qquad (1.2.24)$$

where $m \in C(R_+, K)$.
Then the inequality $m(t_0) \le u_0$ implies that

$$m(t) \le r(t), \quad t \ge t_0. \qquad (1.2.25)$$

Proof We will follow the proof of Theorem 1.2.10, but the inequalities will now be considered with respect to the cone K. It suffices to prove that

$$m(t) \le u(t, \varepsilon), \quad t \in [t_0, T]. \qquad (1.2.26)$$

If the inequality (1.2.26) does not hold, then there exists $t_1 \in (t_0, T]$ such that

$$m(t_1) - u(t_1, \varepsilon) \in \partial K, \quad m(t) - u(t, \varepsilon) \in K^0, \quad t \in [t_0, t_1).$$

It means that there exists $\varphi \in K_0^*$ such that

$$\varphi(m(t_1) - u(t_1, \varepsilon)) = 0.$$

From the quasimonotonicity of the function g it follows that

$$\varphi\{g[t_1, m(t_1)] - g[t_1, u(t_1, \varepsilon)]\} \ge 0.$$

Assuming $w(t) = \varphi(m(t) - u(t, \varepsilon))$, $t \in [t_0, t_1]$, obtain $w(t) > 0$, $t \in [t_0, t_1)$ and $w(t_1) = 0$. Hence $D_-w(t_1) \le 0$, and as a result, we have

$$D_-w(t_1) = \varphi(D_-m(t_1) - u'(t_1, \varepsilon)) > \varphi\{g[t_1, m(t_1)] - g[t_1, u(t_1, \varepsilon)]\} \ge 0,$$

which is a contradiction. Theorem 1.2.13 is proved.

It is clear that the quasimonotonicity of $g(t, u)$ with respect to the cone P does not imply the quasimonotonicity of $g(t, u)$ with respect to the cone Q, even if $P \subset Q$. However, the ordering relationship with respect to the cone P assumes the same ordering relationship with respect to Q if $P \subset Q$. The corollary given below is the result of such observations and is effective in applications.

Corollary 1.2.2 Assume that P, Q are two cones in a space R^n, such that $P \subset Q$. Let the assumptions of Theorem 1.2.13 hold true and $K \equiv P$.

Then the inequality $m(t_0) \leq u_0$ implies that $m(t) \leq r(t)$ at all $t \geq t_0$.

1.2.4   Integral inequalities

Consider a theorem that generalizes Gronwall-Bihari type inequalities.

Theorem 1.2.14 *Let $g \in C(R_+^2, R_+)$, let $g(t, u)$ be a function nondecrescent with respect to u for each $t \in R_+$, and let $r(t)$ be the maximum solution of the system*

$$u' = g(t, u), \quad u(t_0) = u_0, \qquad (1.2.27)$$

existing on the interval $[t_0, \infty)$. Assume that the function $m \in C(R_+, R_+)$ and satisfies the inequality

$$m(t) \leq m(t_0) + \int_{t_0}^{t} g(s, m(t)) \, ds, \quad t \geq t_0. \qquad (1.2.28)$$

Then the condition $m(t_0) \leq u_0$ implies the inequality $m(t) \leq r(t)$ at all $t \geq t_0$.

Proof Assume

$$m(t_0) + \int_{t_0}^{t} g(s, m(t)) ds = v(t),$$

so that $m(t) \leq v(t)$, $m(t_0) = v(t_0)$ and $v' \leq g(t, v)$, proceeding from the fact that the function g is not decrescent with respect to u. Applying Theorem 1.2.10, obtain $v(t) \leq r(t)$, $t \geq t_0$, which completes the proof.

Corollary 1.2.3 Let all the assumptions of Theorem 1.2.14 be correct, expect the inequality (1.2.28) which is replaced by the following:

$$m(t) \leq n(t) + \int_{t_0}^{t} g(s, m(s)) \, ds, \quad t \geq t_0,$$

where $n \in C(R_+, R_+)$.

Then the following inequality holds

$$m(t) \leq n(t) + r(t), \quad t \geq t_0,$$

where $r(t)$ is the maximum solution of the equation

$$u' = g(t, n(t) + u), \quad u(t_0) = 0.$$

1.3 Stability in the Sense of Lyapunov

In its classical statement, the second Lyapunov method combines a number of theorems on stability, asymptotic stability, and instability obtained on the basis of a scalar Lyapunov function and its full derivatives by virtue of motion equations considered in the time-invariant neighborhood of a point $x = 0$. In this section, sufficient conditions for different types of stability of the state $x = 0$ are given in terms of existence of Lyapunov functions which have special properties. Somewhat different versions of those statements were given in the works of Lyapunov [1], Persidsky [1], Grujić, Martynyuk, Ribbens-Pavella [1], Yoshizawa [2], Rao [1], and others.

1.3.1 Lyapunov functions

In this subsection we will consider a system of the form

$$\frac{dx}{dt} = f(t, x), \quad f(t, 0) = 0 \tag{1.3.1}$$

in the range of values (t, x): $t \geq 0$, $\|x\| \leq h$, where $x \in R^n$ and $f \colon R_+ \times R^n \to R^n$.

Comparison functions are used as upper or lower estimates of the function v and its total time derivative. From now on those functions will be denoted by φ, $\phi \colon R_+ \to R_+$. Systematic application of comparison functions is connected with the work of Hahn [1].

Definition 1.3.1 The function φ, $\phi : R_+ \to R_+$, belongs:

(a) to the class $K_{[0,\alpha)}$, $0 < \alpha \leq +\infty$, if it is defined, continuous, and strictly increscent on $[0, \alpha)$ and $\varphi(0) = 0$;

(b) to the class K, if the conditions of Definition 1.3.1 (a) are satisfied at $\alpha = +\infty$, $K = K_{[0,+\infty)}$;

(c) to the class KR, if it belongs to the class K and, in addition, $\varphi(\xi) \to +\infty$ at $\xi \to +\infty$;

(d) to the class $L_{[0,\alpha)}$, if it is defined, continuous, and strictly decrescent on $[0,\alpha)$ and $\lim[\varphi(\zeta)\colon \zeta \to \infty] = 0$;

(e) to the class L, if the conditions of the definition (d) are satisfied at $\alpha = +\infty$, $L = L_{[0,+\infty)}$.

Let φ^{-1} be the inverse function to φ, $\varphi^{-1}[\varphi(\zeta)] \equiv \zeta$. The following properties of comparison functions are known.

Proposition 1.3.1 *If*:

(a) $\varphi \in K$ *and* $\psi \in K$, *then* $\varphi(\psi) \in K$;

(b) $\varphi \in K$ *and* $\sigma \in L$, *then* $\varphi(\sigma) \in L$;

(c) $\varphi \in K_{[0,\alpha]}$ *and* $\varphi(\alpha) = \xi$, *then* $\varphi^{-1} \in K_{[0,\xi]}$;

(d) $\varphi \in K$ *and* $\lim[\varphi(\zeta)\colon \zeta \to +\infty] = \xi$, *then* φ^{-1} *is not defined on* $(\xi, +\infty]$;

(e) $\varphi \in K_{[0,\alpha]}$, $\psi \in K_{[0,\alpha]}$ *and* $\varphi(\zeta) > \psi(\zeta)$ *on* $[0,\alpha]$, *then* $\varphi^{-1}(\zeta) > \psi^{-1}(\zeta)$ *on* $[0,\beta]$, *where* $\beta = \psi(\alpha)$.

Now auxiliary functions will be used, which have the sense of a distance from the origin of coordinates to the current value of solution and play the core role in the direct Lyapunov method (see Grujić, Martynyuk, Ribbens-Pavella [1]).

Definition 1.3.2 The function $v\colon R^n \to R$ is said to be:

(1) positive semidefinite, if there exists a time-invariant neighborhood N, $N \subseteq R^n$, of the point $x = 0$, such that:

   (a) $v(x)$ is continuous on N,

   (b) $v(x)$ is nonnegative on N for all $x \in N$,

   (c) $v(x)$ is equal to zero in the point $x = 0$;

(2) positive semidefinite in the neighborhood S of the point $x = 0$, if the conditions of Definition 1.3.2 (1) are satisfied at $N = S$;

(3) globally positive semidefinite, if the conditions of Definition 1.3.2 (1) are satisfied at $N = R^n$;

(4) negative semidefinite (in the neighborhood S of the point $x = 0$ or in large), if $(-v)$ is positive semidefinite (in the neighborhood S or in large).

Remark 1.3.1 The function v determined by the expression $v(x) = 0$ at all $x \in R^n$ is both positive and negative semidefinite. This indefiniteness can be eliminated by introduction of a strictly positive (negative) semidefinite function.

Definition 1.3.3 The function $v\colon R^n \to R$ is said to be strictly positive (negative) semidefinite if it is positive (negative) semidefinite and there exists $y \in N$ such that $v(y) > 0$ $(v(y) < 0)$.

If the matrix H is strictly positive (negative) semidefinite, then the function $v(x) = x^T H x$ is strictly positive (negative) semidefinite.

Definition 1.3.4 The function $v\colon R^n \to R$ is said to be:

(a) positive definite if there exists a time-invariant neighborhood N, $N \subseteq R^n$, of the point $x = 0$ at which it is positive semidefinite and $v(x) > 0$ at all $(x \neq 0) \in N$;

(b) positive definite in the neighborhood S of the point $x = 0$, if the conditions of Definition 1.3.4 (a) are satisfied at $N = S$;

(c) globally positive definite, if the conditions of Definition 1.3.4 (a) are satisfied at $N = R^n$;

(d) negative definite (in the neighborhood S of the point $x = 0$ or in large) if $(-v)$ is positive definite (in the neighborhood S or in large).

The following statement is known (see Hahn [1]).

Proposition 1.3.2 *For a function v to be positive definite in a neighborhood N of the point $x = 0$, it is necessary and sufficient that a function $\varphi \in K_{[0,\alpha)}$, $\alpha = \sup\{\|x\|\colon x \in N\}$ should exist, such that $v(x) \in C(N)$ and $\varphi(\|x\|) \le v(x)$ at all $x \in N$.*

Definition 1.3.5 The function $v\colon R_+ \times R^n \to R$ is said to be:

(1) positive semidefinite, if there exists a time-invariant connected neighborhood N, $N \subseteq R^n$, of the point $x = 0$, such that:

    (a) v is continuous with respect to $(t, x) \in R_+ \times N$,

    (b) v is nonnegative on N at all $(t, x) \in R_+ \times N$,

    (c) v vanishes at $x = 0$;

(2) positive semidefinite on $R_+ \times S$ if the conditions (a)–(c) of Definition 1.3.5 (1) are satisfied at $N = S$;

(3) globally positive semidefinite, if the conditions (a)–(c) of Definition 1.3.5 (1) are satisfied at $N = R^n$;

(4) negative semidefinite (globally) if $(-v)$ is positive semidefinite (globally);

(5) strictly positive semidefinite if the conditions (a) – (c) of Definition 1.3.5 (1) are satisfied and for each $t \in R_+$ there exists an $y \in N$ such that $v(t,y) > 0$.

Definition 1.3.6 The function $v \colon R \times R^n \to R$ is said to be:

(1) positive definite, if there exists a time-invariant connected neighborhood N, $N \subseteq R^n$, of the point $x = 0$, such that v is positive semidefinite and there exists a positive definite function w on N, $w \colon R^n \to R$, which satisfies the inequality

$$w(x) \leq v(t,x) \quad \text{at all} \quad (t,x) \in R_+ \times N;$$

(2) positive definite on $R_+ \times S$, if all the conditions of Definition 1.3.6 (1) are satisfied at $N = S$;

(3) globally positive definite, if the conditions of Definition 1.3.6 (1) are satisfied at $N = R^n$;

(4) negative definite (globally), if $(-v)$ is positive definite (globally).

The following result is obtained directly from Proposition 1.3.2 and Definition 1.3.4.

Proposition 1.3.3 *For the function $v \colon R_+ \times R^n \to R$ to be positive definite it is necessary and sufficient that there should exist a time-invariant neighborhood N of the point $x = 0$, such that:*

(a) $v(t,x) \in C(R_+ \times N)$;

(b) $v(t,0) = 0$ *at all* $t \in R_+$;

(c) *there exists a function* $\varphi_1 \in K_{[0,\alpha]}$, *where*

$$\alpha = \sup\{\|x\| \colon x \in N\},$$

which satisfies the estimate

$$\varphi_1(\|x\|) \leq v(t,x) \quad \text{at all} \quad (t,x) \in R_+ \times N.$$

Definition 1.3.7 The function $v \colon R \times R^n \to R$ is said to be:

(1) decrescent, if there exist a time-invariant neighborhood N of the point $x = 0$ and a positive definite function w, $w \colon R^n \to R$, such that

$$v(t,x) \leq w(x) \quad \text{at all} \quad (t,x) \in R_+ \times N;$$

(2) decrescent on $R_+ \times S$, if all the conditions of Definition 1.3.7 (1) are satisfied at $N = S$;

(3) globally decrescent on R_+, if all the conditions of Definition 1.3.7 (1) are satisfied at $N = R^n$.

Proposition 1.3.2 and Definition 1.3.7 result in the following statement.

Proposition 1.3.4 *For the function* $v\colon R_+ \times R^n \to R$ *to be decrescent on* $R_+ \times N$, *where* N *is a time-invariant neighborhood of the point* $x = 0$, *it is necessary and sufficient that there should exist a comparison function* $\varphi_2 \in K_{[0,\alpha)}$, $\alpha = \sup\{\|x\|\colon x \in N\}$, *such that*

$$v(t,x) \leq \varphi_2(\|x\|) \quad at\ all \quad (t,x) \in R_+ \times N.$$

Definition 1.3.8 The function $v\colon R_+ \times R^n \to R$ is said to be radially unbounded if at $\|x\| \to +\infty$ $v(t,x) \to +\infty$ at all $t \in R_+$.

It is not difficult to verify the correctness of the above statement (see Hahn [1], Krasovsky [1]).

Proposition 1.3.5 *For a globally positive definite function* v *to be radially unbounded, it is necessary and sufficient that there should exist a function* φ_3 *belonging to the KR-class, such that*

$$v(t,x) \geq \varphi_3(\|x\|) \quad at\ all \quad x \in R^n \quad and \quad t \in R_+.$$

1.3.2 Stability theorems

In the frame of the direct Lyapunov method, the following expressions of full derivative of an auxiliary function v along solutions of the system (1.3.1) are applied.

Let v be a continuous function, $v(t,x) \in C(R_+ \times N)$ and let a solution $\chi(t, t_0, x_0)$ of the system (1.3.1) exist and be defined on $R_+ \times N$. Then for $(t,x) \in R_+ \times N$:

(1) $D^+ v(t,x) = \limsup\{[v[t + \theta, \chi(t + \theta; t, x)] - v(t,x)]\theta^{-1}\colon \theta \to 0^+\}$ is called the upper right-hand Dini derivative of the function v along the solution $\chi(t, t_0, x_0)$;

(2) $D_+ v(t,x) = \liminf\{[v[t+\theta, \chi(t+\theta; t, x)] - v(t,x)]\theta^{-1}\colon \theta \to 0^+\}$ is called the lower right-hand Dini derivative of the function v along the solution $\chi(t, t_0, x_0)$;

(3) $D^- v(t,x) = \limsup\{[v[t + \theta, \chi(t + \theta; t, x)] - v(t,x)]\theta^{-1}\colon \theta \to 0^-\}$ is called the upper left-hand Dini derivative of the function v along the solution $\chi(t, t_0, x_0)$;

(4) $D_-v(t,x) = \limsup\{[v[t+\theta, \chi(t+\theta; t,x)] - v(t,x)]\theta^{-1} : \theta \to 0^-\}$ is called the lower left-hand Dini derivative of the function v along the solution $\chi(t, t_0, x_0)$;

(5) The function v has a Eulerian derivative

$$\dot{v}(t,x) = \frac{d}{dt} v(t,x)$$

along the solution $\chi(t, t_0, x_0)$, if

$$D^+v(t,x) = D_+v(t,x) = D^-v(t,x) = D_-v(t,x) = Dv(t,x),$$

and then

$$\dot{v}(t,x) = Dv(t,x).$$

If v is differentiable with respect to $(t,x) \in R_+ \times N$, then

$$\dot{v} = \frac{\partial v}{\partial t} + (\operatorname{grad} v)f(t,x),$$

where

$$\operatorname{grad} v = \left(\frac{\partial v}{\partial x_1}, \frac{\partial v}{\partial x_2}, \ldots, \frac{\partial v}{\partial x_n} \right)^{\mathrm{T}}.$$

The effective application of the upper right-hand Dini derivative within the limits of the direct Lyapunov method is based on the following result obtained by Yoshizawa [2], which secures the evaluation of D^+v without direct use of solutions of the system (1.3.1).

Theorem 1.3.1 *Let the function v be continuous and locally Lipshitz with respect to x in the product $R_+ \times S$ and let S be an open set. Then along a solution χ of the system (1.3.1)*

$$D^+v(t,x) = \limsup \left\{ \frac{v[t+\theta, x+\theta f(t,x)] - v(t,x)}{\theta} : \theta \to 0^+ \right\}$$

at $(t,x) \in R_+ \times S$.

Further the symbol D^*v will mean that it is admissible to use both D^+v and D_+v.

Theorem 1.3.2 *Let the vector function f in the system (1.3.1) be continuous on $R_+ \times N$. If there exist:*

(1) *an open time-invariant connected neighborhood S of the point $x = 0$;*

(2) *a positive definite function v on S such that*

$$D^+v(t,x) \le 0 \quad \text{at all} \quad (t,x) \in R_+ \times S,$$

then the state $x = 0$ of the system (1.3.1) is stable.

Proof Since the function $v(t,x)$ is positive definite, there exists a function φ_1 belonging to the K-class, such that

$$\varphi_1(\|x\|) \leq v(t,x) \text{ at all } (t,x) \in R_+ \times B(\rho),$$

where $B(\rho) \subset R^n$ is an open connected domain in R^n. Now, at any $0 < \varepsilon < \rho$ and $t_0 \in R_+$ choose $\delta = \delta(t_0, \varepsilon)$ so that the condition $\|x_0\| \leq \delta$ would imply $v(t_0, x_0) < \varphi_1(\varepsilon)$. This is possible, since $v(t,x)$ is a continuous function and $v(t_0, 0) = 0$. We will show that for any solution $x(t; t_0, x_0) = x_0$ at $t = t_0$ under the condition $\|x_0\| < \delta$ the inequality $\|x(t; t_0, x_0)\| < \varepsilon$ holds at all $t \geq t_0$. If this is not so, then one can find a $t^* > t_0$, for which

$$\|x(t^*; t_0, x_0)\| = \varepsilon \text{ at } \|x_0\| < \delta.$$

From condition (2) of the theorem it follows that

$$v(t, x(t; t_0, x_0) \leq v(t_0, x_0)) \quad \text{at all} \quad t \geq t_0.$$

Hence at $t = t^*$ obtain

$$\varphi_1(\varepsilon) = \varphi_1(\|x(t^*; t_0, x_0)\|) \leq v(t^*, x(t^*, t_0, x_0)) \leq v(t_0, x_0) < \varphi_1(\varepsilon).$$

From the obtained contradiction it follows that at a point of time t^* the inequality $\|x(t, t_0, x_0)\| < \varepsilon$ holds at $\|x_0\| < \delta$, which completes the proof of the theorem.

Example 1.3.1 Consider the scalar equation

$$\frac{dx}{dt} = (\sin \log t + \cos \log t - 1{,}25)x. \tag{1.3.2}$$

Choose a Lyapunov function in the form

$$v(t,x) = x^2 \exp[2(1{,}25 - \sin \log t)t].$$

This function is positive definite, but nondecrescent. Since $D^+ v(t,x)|_{(1.3.2)} = 0$, the zero solution of the equation (1.3.2) is stable.

Theorem 1.3.3 *Let the vector function f in the system (1.3.1) be continuous on $R_+ \times N$. If there exist:*

(1) *an open time-invariant connected neighborhood S of the point $x = 0$;*

(2) *a decrescent positive definite function v on S such that*

$$D^+ v(t,x) \leq 0 \text{ at all } (t,x) \in R_+ \times S,$$

then the condition $x = 0$ of the system (1.3.1) is uniformly stable.

Proof Since the function $v(t, x)$ is positive definite and decrescent, there exist functions φ_1, φ_2 belonging to the K-class, such that

$$\varphi_1(\|x\|) \le v(t, x) \le \varphi_2(\|x\|) \quad \text{at all} \quad (t, x) \in R_+ \times B(\rho).$$

For any $0 < \varepsilon < \rho$ choose $\delta = \delta(\varepsilon) > 0$ so that $\varphi_2(\delta) < \varphi_1(\varepsilon)$. Show that under the conditions of Theorem 1.3.3 the zero solution of the system (1.3.1) is uniformly stable, that is, if $t^* \ge t_0$ and $\|x(t^*)\| \le \delta$, then $\|x(t; t^*, x_0)\| < \varepsilon$ at all $t \ge t^*$. If this is not so, then there exists $\hat{t} \ge t^*$ such that at $t^* \ge t_0$ and $\|x(t^*)\| \le \delta$ the relation $\|x(\hat{t}; t^*, x_0)\| = \varepsilon$ holds. Like in the proof of Theorem 1.3.2, obtain

$$\varphi_1(\varepsilon) = \varphi_1(\|x(\hat{t})\|) \le v(\hat{t}, x(\hat{t}; t^*, x_0)) \le v(t^*, x(t^*))$$
$$\le \varphi_2(\|x(t^*)\|) \le \varphi_2(\delta) < \varphi_1(\varepsilon).$$

The obtained contradiction shows that $\hat{t} \notin R_+$ and at $\|x(t^*)\| \le \delta$ the estimate $\|x(t; t^*, x_0)\| < \varepsilon$ will hold at all $t \ge t^*$. Theorem 1.3.3 is proved.

Theorem 1.3.4 *Let the vector function f in the system (1.3.1) be continuous on $R_+ \times N$ and bounded. If there exist:*

(1) *an open time-invariant connected neighborhood S of the point $x = 0$;*

(2) *a function $v(t, x)$ positive definite on S and a function ψ belonging to the K-class, such that*

$$D^* v(t, x) \le -\psi(\|x\|) \quad \text{at all} \quad (t, x) \in R_+ \times S,$$

then the condition $x = 0$ of the system (1.3.1) is asymptotically stable.

Proof From condition (2) of the theorem it follows that $D^* v(t, x) \le 0$ at all $(t, x) \in R_+ \times S$, and this condition together with condition (1) secures the stability of the state $x = 0$ of the system (1.3.1), that is, for any $0 < \varepsilon < \rho$ and $t_0 \in R_+$ there exists $\delta = \delta(t_0, \varepsilon) > 0$ such that the condition $\|x_0\| < \delta$ implies that $\|x(t; t_0, x_0)\| < \varepsilon$ at all $t \ge t_0$.

We will show that the state $x = 0$ of the system (1.3.1) is asymptotically stable. Let this not be so and let t_0 and $x_0 \in S$ be such that for some η, $0 < \eta < \varepsilon$, and a divergent sequence $\{t_n\} \in R_+$ the equality $\|x(t_n; t_0, x_0)\| = \eta$ would hold. Since f is bounded on $R_+ \times N$, there exists a constant $M > 0$ such that $\|f(t, x)\| < M$ on $R_+ \times N$. In this case, for $t \ge t_0$ obtain the estimate

$$\|x(t; t_0, x_0) - x(t_n; t_0, x_0)\| \le M|t - t_n|, \quad n = 1, 2, \ldots,$$

and then

$$\|x(t; t_0, x_0)\| \ge \frac{1}{2}\eta \quad \text{at} \quad t \in I_n = \left(t_n - \frac{\eta}{2M}, t_n + \frac{\eta}{2M}\right).$$

Assume that the intervals I_n disjoint, that is, $I_n \cap I_{n+1} = \varnothing$, $n = 1, 2, \ldots$, and choose $t_1 > t_0 + \dfrac{\eta}{2M}$. From condition (2) of Theorem 1.3.4 it follows that at $t \geq t_0$ and $\dfrac{1}{2}\eta \leq \|x\| < \rho$ there exist constants a and b such that $b \leq v(t, x)$ and $D^* v(t, x) \leq -a$. Hence, at $t \in \left[t_0, t_n + \dfrac{\eta}{2M} \right]$, obtain

$$0 < b \leq v\left(t_n + \frac{\eta}{2M}, x\left(t_n + \frac{\eta}{2M}, t_0, x_0 \right) \right) - v(t_0, x_0)$$

$$\leq -a\left(t_n - t_0 + \frac{\eta}{2M} \right) \leq -a\frac{\eta}{2M} - \frac{an\eta}{M} < -a\frac{\eta}{M}n \to -\infty$$

at $n \to \infty$.

The obtained contradiction proves that the state $x = 0$ of the system (1.3.1) is asymptotically stable.

Theorem 1.3.5 *Let the vector function f in the system (1.3.1) be continuous on $R_+ \times N$. If there exist:*

(1) *an open time-invariant connected neighborhood S of the point $x = 0$;*

(2) *a decrescent positive definite function v on S;*

(3) *a positive definite function ψ on S such that*

$$D^* v(t, x) < -\psi(x) \quad \text{at all} \quad (t, x) \in R_+ \times S,$$

then the state $x = 0$ of the system (1.3.1) is uniformly asymptotically stable.

Proof Let $0 < \varepsilon < \rho$ be specified. Condition (2) of Theorem 1.3.5 implies the existence of functions φ_1, φ_2 belonging to the K-class, such that

$$\varphi_1(\|x\|) \leq v(t, x) \leq \varphi_2(\|x\|) \quad \text{at all} \quad (t, x) \in R_+ \times S.$$

From condition (3) it follows that $D^* v(t, x) \leq 0$, then the state $x = 0$ of the system (1.3.1) is uniformly stable. Here for any $\varepsilon > 0$ there exists a $\delta = \delta(\varepsilon) > 0$ such that for any solution $x(t) = x(t; t_0, x_0)$ of the system (1.3.1) the conditions $\|x(t_1)\| \leq \delta$, $t_1 \geq t_0$ imply that $\|x(t)\| < \varepsilon$ at all $t \geq t_1$. From condition (3) obtain

$$v(t, x(t)) \leq v(t_1, x(t_1)) - \int_{t_1}^{t} \psi(\|x(s)\|)\, ds. \tag{1.3.3}$$

Let $0 < \eta < \rho$ be specified, choose $\delta_0 = \delta(\rho)$ and $T(\eta) = \dfrac{\varphi_2(\delta_0)}{\psi(\delta(\eta))}$. Show that $\|x(t)\| < \delta(\eta)$ at all $t \in [t_1, t_1 + T]$, as soon as $t_1 \geq t_0$ and $\|x(t_1)\| < \delta_0$. Assume that this is not so. Then there exists a $t \in [t_1, t_1 + T]$ for which

$$\|x(t)\| \geq \delta(\eta). \tag{1.3.4}$$

From the conditions (1.3.3), (1.3.4) at $t \in [t_1, t_1 + T]$ obtain

$$v(t, x(t)) \leq v(t_1, x_1) - \psi(\delta(\eta))(t - t_1) \leq \varphi_2(\delta_0) - \psi(\delta(\eta))(t - t_1).$$

Hence for $t = t_1 + T$ obtain

$$0 < \varphi_1(\delta(\eta)) \leq v(t_1 + T, x(t_1 + T)) \leq \varphi_2(\delta_0) - \psi(\delta(\eta))T(\eta) = 0.$$

This contradicts the inequality (1.3.4) and therefore there exists $t_1 \leq t_2 \leq t_1 + T$, for which $\|x(t_2; t_1, x_1)\| < \delta(\eta)$.

Thus, from the uniform stability of the state $x = 0$ it follows that $\|x(t)\| < \eta$ at all $t \geq t_2$, in particular at $t \geq t \geq t_1 + T$. Consequently, $\|x(t)\| < \eta$ at all $t \geq t_1 + T$, as soon as $t_1 \geq t_0$ and $\|x(t_1)\| \leq \delta_0$. This proves that the state $x = 0$ of the system (1.3.1) is uniformly asymptotically stable, since T depends on η only.

Example 1.3.2 Consider the scalar equation

$$\frac{dx}{dt} = (t \sin t - 2t)x, \quad x(t_0) = x_0. \tag{1.3.5}$$

Take a Lyapunov function in the form

$$v(t, x) = x^2 \exp \left[\int_0^t (2u - u \sin u) \, du \right].$$

This function is positive definite, but nondecrescent. A simple calculation results in the estimate

$$D^+ v(t, x)|_{(1.3.5)} \leq -\alpha v(t, x) \quad \text{at all} \quad t > \alpha \geq 0.$$

Therefore, the zero solution of the equation (1.3.5) is asymptotically stable, however not uniformly.

The property of exponential stability of the zero solution of the generating system (1.3.1) is determined by the following statement.

Definition 1.3.9 The state $x = 0$ of the system (1.3.1) is said to be exponentially stable, if for any solution $x(t)$ of this system in the domain $t \geq t_0$, $x \in B(\rho)$ the following inequlity holds:

$$\|x(t; t_0, x_0)\| \leq a\|x_0\| \exp(-\lambda(t - t_0)),$$

where $a > 0$, $\lambda > 0$, $t_0 \geq 0$. The constants a and λ may depend on B.

Definition 1.3.10 The comparison functions φ_1, φ_2 belonging to the K-class have the same order of growth, if there exist constants α_i, β_i, $i = 1, 2$, such that

$$\alpha_i \varphi_i(r) \leq \varphi_j(r) \leq \beta_i \varphi_i(r), \quad i, j = 1, 2.$$

Theorem 1.3.6 *Let the vector function f in the system (1.3.1) be continuous on $R_+ \times N$ and let there exist:*

(1) *a time-invariant neighborhood S of the state $x = 0$;*

(2) *a function $v(t, x)$ locally Lipshitz with respect to x, comparison functions φ_1, φ_2 belonging to the K-class, having the same order of growth, and constants $a > 0$, $r_1 > 0$ such that*

$$a\|x\|^{r_1} \leq v(t, x) \leq \varphi_1(\|x\|)$$

at all $(t, x) \in R_+ \times S$;

(3) *at all $(t, x) \in R_+ \times S$ the following estimate holds $D^*v(t, x) \leq -\varphi_2(\|x\|)$.*

Then the state $x = 0$ of the system (1.3.1) is exponentially stable.

Proof Taking into account that the comparison functions φ_1 and φ_2 have the same order of growth, it is possible to indicate constants $\alpha_1, \beta_1 > 0$ such that

$$\alpha_1\varphi_1(r) \leq \varphi_2(r) \leq \beta_1\varphi_1(r).$$

Therefore, condition (3) of the theorem is reducible to the form

$$D^*v(t, x) \leq -\alpha_1 v(t, x) \quad \text{at all} \quad (t, x) \in R_+ \times S.$$

Hence obtain

$$v(t, x(t)) \leq v(t_0, x_0)\exp(-\alpha_1(t - t_0)) \quad \text{at all} \quad t \geq t_0.$$

From the above estimate and condition (2) of the theorem it follows that

$$\|x(t; t_0, x_0)\| \leq a^{-1/r_1}\varphi_1^{1/r_1}(\|x_0\|)\exp\left(-\frac{\alpha_1}{r_1}(t - t_0)\right) \tag{1.3.6}$$

at all $t \geq t_0$.

Denote $\lambda = \dfrac{\alpha_1}{r_1}$. Then the estimate (1.3.6) is equivalent to the determination of the exponential stability of the state $x = 0$ of the system (1.3.1). In addition, for any $\varepsilon > 0$ it is possible to choose $\delta(\varepsilon) = \varphi_1^{-1}(a\varepsilon^{r_1})$ so that as soon as $\|x_0\| < \delta(\varepsilon)$, $t_0 \geq 0$, then

$$\|x(t; t_0, x_0)\| \leq \varepsilon\exp(-\lambda(t - t_0)) \quad \text{at all} \quad t \geq 0.$$

The theorem is proved.

Theorem 1.3.7 *Let the vector function f in the system (1.3.1) be continuous on $R_+ \times N$. If there exist a function $v(t, x)$ locally Lipshitz with respect to x and comparison functions φ_2, ψ belonging to the K-class, such that:*

(1) *at all $(t, x) \in R_+ \times B(\rho)$ the estimate $|v(t, x)| \leq \varphi_2(\|x\|)$ holds;*

(2) *for any $\delta > 0$ and any $t_0 > 0$ there exists x_0, $\|x_0\| \leq \delta$ such that $v(t, x) < 0$;*

(3) *at all* $(t, x) \in R_+ \times B(\rho)$ *the following estimate holds* $D^*v(t, x) \le -\psi(\|x\|)$,

then the state $x = 0$ *of the system (1.3.1) is unstable.*

Proof Let $0 < \varepsilon < \rho$ be specified. Assume that the state $x = 0$ of the system (1.3.1) is stable. Then for any $\varepsilon > 0$ there exists $\delta = \delta(\varepsilon) > 0$ such that the condition $\|x(t_0)\| \le \delta$ implies the estimate $\|x(t)\| < \varepsilon$ at all $t \ge t_0$. Choose x_0 so that $\|x_0\| \le \delta$ and $v(t_0, x_0) < 0$. According to the assumption, at $\|x_0\| < \delta$ the inequality $\|x(t)\| < \varepsilon$ holds at all $t \ge t_0$. Condition (1) of the theorem implies that

$$|v(t, x(t))| \le \varphi_2(\|x(t)\|) < \varphi_2(\varepsilon) \quad \text{at any} \quad t \ge t_0.$$

From condition (3) of the theorem it follows that the function $v(t, x(t))$ is decrescent along any solution $x(t)$ of the system (1.3.1); therefore, for any $t \ge t_0$ the following estimate holds:

$$v(t, x(t)) \le v(t_0, x_0) < 0.$$

Hence, follows that $|v(t, x(t))| \ge |v(t_0, x(t_0))|$. Condition (3) of Theorem 1.3.7 implies that

$$v(t, x(t)) \le v(t_0, x_0) - \int_{t_0}^{t} \psi(\|x(s)\|)ds. \tag{1.3.7}$$

According to condition (1) of the theorem, we obtain $\|x(t)\| \ge \varphi_2^{-1}(|v(t_0, x_0)|)$; therefore, $\psi(\|x(t)\|) \ge \psi(\varphi_2^{-1}(|v(t_0, x_0)|))$. Taking this inequality into account, transform (1.3.7) to the form

$$v(t, x(t)) \le v(t_0, x_0) - \psi(\varphi_2^{-1}(|v(t_0, x_0)|))(t - t_0).$$

Hence it follows that $\lim_{t \to \infty} v(t, x(t)) = -\infty$, which contradicts the condition $|v(t, x(t))| < \varphi_2(\varepsilon)$ at all $t \ge t_0$. This proves the instability of the state $x = 0$ of the system (1.3.1).

The universality of stability theorems is determined by the respective converse theorems. One of the first theorems in this line of research is Persidsky's theorem [1] on the existence of a Lyapunov function in case of stability of the state $x = 0$ of the system (1.3.1). The works of Krasovsky [1], Hahn [1], Zubov [2], and others contain results concerned with the inversion of Lyapunov theorems. Note that the known proofs of converse theorems are based on functions that contain an estimate of solution of a differential equation which is unknown as a rule. Among these functions are:

(1) $v(t, x) = (1 + e^{-t})\|y(t_0; t, x)\|^2$ in the domain $t \ge t_0$, $\|x\| < \rho < +\infty$. Here $y(t; t_0, y_0)$ is a solution of the system

$$\frac{dy}{dt} = Y(t, y), \quad Y(t, y) = f(t, x)\phi(y),$$

where $\phi(y) = \begin{cases} 1 & \text{at } \|y\| \leq \rho; \\ 0 & \text{at } \|y\| > \rho; \end{cases}$

(2) $v(t,x) = \sup_{\tau \geq 0} \|x(t+\tau; t, x)\|$;

(3) $v(t,x) = \sup_{\tau \geq 0} \|x(t+\tau; t, x)\| \exp(\alpha q \tau)$, where $\alpha > 0$, $q \in (0,1)$;

(4) $v(t,x) = \sup_{\sigma \geq 0} G(\|x(t+\sigma; t, x)\|) \dfrac{1+\alpha\sigma}{1+\sigma}$, where $G(r)$ is a scalar function
with the properties $G(0) = 0$, $G'(0) = 0$, $G'(0) > 0$, $G''(r) > 0$ and
$\alpha > 1$;

(5) $v(t,x) = \sup_{\sigma \geq 0} \varphi(\|x(t+\sigma; t, x)\|)$, where φ belongs to K-class.

It is clear that functions of the form (1)–(5) have no practical use, but they show that under a certain type of stability of the state $x = 0$ of the system (1.3.1) there exists a Lyapunov function with the respective properties.

1.4   Comparison Principle

The theorems of the comparison method given in this section are based on the scalar Lyapunov function for a generating system and the theory of differential inequalities. For their further application it is sufficient to consider the case of continuous solutions both in the initial system and in the comparison system.

Consider the system of differential equations

$$\frac{dx}{dt} = f(t,x), \quad x(t_0) = x_0, \tag{1.4.1}$$

where $x \in R^n$, $t \in R_+$, $f \in C(R_+ \times R^n, R^n)$. Together with the system (1.4.1) we consider the Lyapunov function $v(t,x)$ and its full derivative $D^*v(t,x)$ along solutions of the system (1.4.1).

Formulate the main theorems of the comparison method.

Theorem 1.4.1 *Let the function* $v \in C(R_+ \times R^n, R_+)$ *be locally Lipschitz with respect to* x. *Assume that the function* $D^+v(t,x)$ *satisfies the inequality*

$$D^+v(t,x)|_{(1.4.1)} \leq g(t, v(t,x)) \quad \text{at all} \quad (t,x) \in R_+ \times R^n, \tag{1.4.2}$$

where $g \in C(R_+^2, R)$. *Let* $r(t) = r(t, t_0, u_0)$ *be the maximum solution of the scalar differential equation*

$$\frac{du}{dt} = g(t,u), \quad u(t_0) = u_0, \tag{1.4.3}$$

existing at all $t \geq t_0$.

    Then the inequality $v(t_0, x_0) \leq u_0$ implies the estimate

$$v(t, x(t)) \leq r(t) \quad at\ all \quad t \geq t_0, \tag{1.4.4}$$

where $x(t) = x(t, t_0, x_0)$ is any solution of the system (1.4.1), which exists at $t \geq t_0$.

    Proof Let $x(t)$ be any solution of the system (1.4.1), existing at $t \geq t_0$, such that $v(t_0, x_0) \leq u_0$. Define the function $m(t) = v(t, x(t))$. For any $h > 0$ obtain

$$m(t + h) - m(t) = v(t + h, x(t + h)) - v(t + h, x(t) + hf(t, x(t)))$$
$$+ v(t + h, x(t) + hf(t, x(t))) - v(t, x(t)).$$

Since the function $v(t, x)$ is locally Lipschitz with respect to x, obtain the differential inequality

$$D^+ m(t) \leq g(t, m(t)), \quad m(t_0) \leq u_0, \tag{1.4.5}$$

and now, in view of Theorem 1.2.10, arrive at the desired result: the estimate (1.4.4).

    Corollary 1.4.1 If in Theorem 1.4.1 we assume that $g(t, u) \equiv 0$, then the function $v(t, x(t))$ is not increscent with respect to t and $v(t, x(t)) \leq v(t_0, x_0)$ at all $t \geq t_0$.

    The next theorem is important at application of several Lyapunov functions.

    Theorem 1.4.2 *Let the vector function $v \in C(R_+ \times R^n, R^m)$, $m \leq n$, be locally Lipschitz with respect to x. Assume that*

$$D^+ v(t, x) \leq g(t, v(t, x)), \quad (t, x) \in R_+ \times R^n,$$

where $g \in C(R_+ \times R_+^m, R^m)$ and $g(t, u)$ is a function quasimonotone nondecrescent with respect to u. Let $r(t) = r(t, t_0, u_0)$ be the maximum solution of the system

$$\frac{du}{dt} = g(t, u), \quad u(t_0) = u_0 \geq 0, \tag{1.4.6}$$

existing at $t \geq t_0$, and let $x(t) = x(t, t_0, x_0)$ be any solution of the system (1.4.1), existing at $t \geq t_0$.
    Then the inequality $v(t_0, x_0) \leq u_0$ implies the estimate

$$v(t, x(t)) \leq r(t) \quad at\ all \quad t \geq t_0. \tag{1.4.7}$$

    Recall that the inequalities between vectors in (1.4.7) are understood componentwise.

Theorem 1.4.2 is a special case of the next theorem connected with cone-valued Lyapunov functions.

Theorem 1.4.3 *Assume that $v \in C(R_+ \times R^n, K)$, the function $v(t,x)$ is locally Lipschitz with respect to x relative to the cone $K \subset R^m$ and at $(t,x) \in R_+ \times R^n$ the following inequality is true:*

$$D^+ v(t,x) \underset{K}{\leqq} g(t, v(t,x)).$$

Let $g \in C(R_+ \times K, R^m)$, here the function $g(t,u)$ is not quasimonoton decrescent with respect to u with respect to K and $r(t) = r(t, t_0, u_0)$ is the maximum solution of the equation (1.4.6), which exists at $t \geq t_0$.

Then any solution $x(t) = x(t, t_0, x_0)$ of the system (1.4.1), existing at $t \geq t_0$, satisfies the estimate

$$v(t, x(t)) \underset{K}{\leqq} r(t) \quad at \ all \quad t \geq t_0$$

provided that $v(t_0, x_0) \underset{K}{\leqq} u_0$.

Proof Following the proof of Theorem 1.4.1, upon necessary changes one can easily obtain the differential inequality

$$D^+ m(t) \underset{K}{\leqq} g(t, m(t)), \quad m(t_0) \underset{K}{\leqq} u_0, \quad t \geq t_0.$$

Now, using Theorem 1.2.13, complete the proof of Theorem 1.4.3.

The next theorem, which is an alternate version of Theorem 1.4.3, is more accessible for applications. Its proof follows from Corollary 1.2.2.

Theorem 1.4.4 *Let P and Q be two cones in R^m such that $P \subset Q$. Assume that $v \in C(R_+ \times R^n, Q)$ and the function $v(t,x)$ satisfies the local Lipschitz condition with respect to x relative to the cone P and*

$$D^+ v(t,x) \underset{P}{\leqq} g(t, v(t,x)), \quad (t,x) \in R_+ \times R^n.$$

Now assume that $g \in C(R_+ \times Q, R^m)$, the function $g(t,u)$ is quasimonotone nondecrescent with respect to u relative to the cone P and $x(t) = x(t, t_0, x_0)$ is any solution of the system (1.3.3), existing at $t \geq t_0$, such that $v(t_0, x_0) \underset{P}{\leqq} u_0$. Then

$$v(t, x(t)) \underset{Q}{\leqq} r(t) \quad at \ all \quad t \geq t_0, \tag{1.4.8}$$

where $r(t) = r(t, t_0, u_0)$ is the maximum solution of the equation (1.4.6) with respect to the cone P.

In particular, if $Q = R_+^m$, then the inequality (1.4.8) implies the componentwise estimate $v(t, x(t)) \leq r(t)$, $t \geq t_0$.

Remark 1.4.1 In all of the above comparison theorems the derivative $D^+ v(t,x)$ was estimated by the functions $g(t, v(t,x))$ only. However, in certain cases it is more natural to estimate the derivative $D^+ v(t,x)$ by the functions $g(t, x, v(t,x))$. Obviously, in this case the statements of the theorems would be more complicated.

1.5   Stability of Systems with a Small Parameter

Consider the system of differential equations with a small parameter

$$\frac{dy}{dt} = Y(t, y, \mu), \quad y(t_0) = y_0 \tag{1.5.1}$$

It is assumed that the right-hand part of the system (1.5.1) has the continuity property   and satisfies the conditions   of the existence and   uniqueness of solutions. In addition, assume that $Y(t, 0, \mu) \not\equiv 0$ at all $t \in R_+$, $\mu \in M$. Denote $\mu = z$ and consider the system extended to (1.5.1)

$$\begin{aligned}
\frac{dy}{dt} &= Y(t, y, z), \quad y(t_0) = y_0, \\
\frac{dz}{dt} &= 0.
\end{aligned} \tag{1.5.2}$$

It is not difficult to show that from the properties of solutions of the system (1.5.2) one can obtain the conclusion about the stability of the system (1.5.1).

Example 1.5.1 (see Duboshin [2])   Let the following system of equations be specified:

$$\begin{aligned}
\frac{dx}{dt} &= -\mu x - y - \mu(1 - \mu), \\
\frac{dy}{dt} &= x - \mu y - \mu(1 - \mu),
\end{aligned} \tag{1.5.3}$$

where $\mu \in (0, 1)$, and let a problem of the stability of the zero solution of the system

$$\begin{aligned}
\frac{dx}{dt} &= -y, \\
\frac{dy}{dt} &= x,
\end{aligned} \tag{1.5.4}$$

be considered, which is obtained from the system (1.5.3) at $\mu = 0$. Assuming $\mu = z$, for the system (1.5.3) obtain

$$\begin{aligned}
\frac{dx}{dt} &= y - z - xz + z^2, \\
\frac{dy}{dt} &= x - z - yz - z^2, \\
\frac{dz}{dt} &= 0.
\end{aligned} \tag{1.5.5}$$

For the system (1.5.5) the full derivative of the function

$$v(x, y, z) = (x - z)^2 + (y + z)^2 + z^2 \tag{1.5.6}$$

along solutions of the system (1.5.5) has the form

$$Dv(t,x,z)|_{(1.5.5)} = -2z[(x-z)^2 + (y+z)^2]. \qquad (1.5.7)$$

By implication of the problem $z \in (0,1)$ and therefore (1.5.7) is a negative function with a fixed sign. Since the function (1.5.6) is positive definite, the solution $x = y = 0$ of the system (1.5.4) is stable for a parameter $\mu \in (0,1)$. It is easy to show that for any $\varepsilon > 0$ it is sufficient to choose $\delta = \delta(\varepsilon) = 2\alpha\varepsilon$, where $\alpha = \left(\dfrac{2-\sqrt{3}}{2}\right)^{1/2}$ and $\mu < \alpha\varepsilon$, so that at

$$|x_0| \le 2\alpha\varepsilon, \quad |y_0| \le 2\alpha\varepsilon, \quad \mu < \alpha\varepsilon$$

for any $t \ge 0$ the following estimates should hold:

$$|x(t,t_0,x_0)| < \varepsilon \quad \text{and} \quad |y(t,t_0,x_0)| < \varepsilon.$$

1.5.1   States of equilibrium

For the system (1.5.1) introduce the following definition (cf. Gruijic, Martynyuk, Ribbens-Pavella [1]).

Definition 1.5.1 The state y^* of the system (1.5.1) is the state of equilibrium if

$$y(t;t_0,y^*,\mu) = y^* \quad \text{at all} \quad t \in R_+, \quad t_0 \ge 0, \quad (\mu \ne 0) \in M, \qquad (1.5.8)$$

where $y(t;t_0,y^*,\mu)$ is the motion of the system (1.4.1) at a point of time $t \in R_+$, if and only if $y(t_0;t_0,y^*,\mu) \equiv y_0$.

Proposition 1.5.1 *For $y^* \in R^n$ to be the equilibrium of the system (1.5.1) it is necessary and sufficient that at $(\mu \ne 0) \in M$:*

(1) *for any $t_0 \ge 0$ the solution $y(t;t_0,y_0^*,\mu)$ of the system (1.5.1) should be unique, determined for all $t \in R_+$;*

(2) $Y(t_0,y^*,\mu) = 0, \quad t \in R_+, \quad t_0 \ge 0.$

Proof. The necessity. Conditions (1) and (2) are necessary in view of Definition 1.5.1 [the relation (1.5.8)].

The sufficiency. If $y^* \in R^n$ satisfies condition (2), then $y(t,\mu) = y^*$ at all $t \in R_+$, at all $t_0 \ge 0$ and $(\mu \ne 0) \in M$ so that

$$\frac{d}{dt}y(t,\mu) = 0 = Y(t,y^*,\mu) = Y(t;y(t;\mu),\mu) \qquad (1.5.9)$$

at all $t \in R_+$, $t_0 \ge 0$, $(\mu \ne 0) \in M$.

Hence $y(t;t_0,y^*,\mu) = y^*$ is a solution of the system (1.5.1) at (t_0,y^*,μ),

which is unique at all $t_0 \geq 0$ and $(\mu \neq 0) \in M$. Consequently, the relation (1.5.8) is correct.

Let $y(t; t_0, y_0, \mu)$, $y(t_0; t_0, y_0, \mu) \equiv x_0$, be the motion of the system (1.5.1) and let $y_p(t; t_0, y_0, \mu)$ be a nonperturbed motion.

From the physical point of view, it is a nonperturbed motion that should be realized in the system; from the mathematical point of view, this means that the functions describing the nonperturbed motion are the solution of the system (1.5.1), that is,

$$\frac{d}{dt} y_p(t; t_0, y_0, \mu) \equiv Y(t; y_p(\cdot), \mu). \tag{1.5.10}$$

1.5.2   Definitions of stability

Definition 1.5.2 The equilibrium y^* of the system (1.5.1):

(1) is μ-stable, if and only if for any $t_0 \geq 0$ and every $\varepsilon > 0$ there exist $\delta = \delta(t_0, \varepsilon) > 0$ and $\mu_1(t_0, \varepsilon) > 0$ such that at $\|y_0 - y^*\| < \delta$ the following inequality would hold

$$\|y(t; t_0, y_0, \mu) - y^*\| < \varepsilon \quad \text{at all} \quad t \geq t_0 \quad \text{and} \quad \mu < \mu_1;$$

(2) is uniformly μ-stable, if and only if the conditions of the Definition 1.5.2 (1) are satisfied and at every $\varepsilon > 0$ the respective maximum value of δ in Definition 1.5.2 (1) does not depend on t_0.

Definition 1.5.3 The equilibrium y^* of the system (1.5.1) is μ-attracting, if and only if for any $t_0 \geq 0$ there exists $\Delta(t_0) > 0$ and for any $\rho \in (0, +\infty)$ there exist $\tau(t_0, y_0, \rho) \in [0, +\infty)$ and $\mu_2(t_0, \rho) > 0$ such that at $\|y_0 - y^*\| < \Delta(t_0)$ $\|y(t; t_0, y_0, \mu) - y^*\| < \rho$ holds at all $t \geq t_0 + \tau$ and $\mu < \mu_2$.

Calculate the value $\mu_0 = \min(\mu_1, \mu_2)$. On the basis of the definitions of μ-stability and μ-attraction, the definitions of asymptotic μ-stability are stated as follows.

Definition 1.5.4 The state of equilibrium y^* of the system (1.5.1) at $\mu < \mu_0$ is:

(1) asymptotically μ-stable, if and only if it is μ-stable and μ-attracting;

(2) uniformly asymptotically μ-stable, if and only if it is uniformly μ-stable and uniformly μ-attracting.

The above definitions and terminology are similar to the known systems of definitions used in the literature (see Hahn [1], Yoshizawa [2], etc.). However, one should bear in mind that the system (1.5.1), generally speaking, may have no zero solution, and therefore, unlike classical definitions characterizing a specific solution of a system of differential equations under consideration, the property of μ-stability characterizes the local behavior of a one-parameter family of systems of the form (1.5.1) depending on the numeric parameter μ.

1.6 Comments and References

Differential equations containing a small parameter as models of real processes and phenomena in the engineering and technological areas have been applied for a long time (see Poincaré [2], Krylov, Bogolyubov [2], Bogolyubov, Mitropolsky [1], Stocker [1], Hayashi [1], Nayfeh, Mook [1] et al.).

Among the examples where systems of this kind have been of use, one should note the problem of the loss of stability by a thin-shelled structure under the action of wind and the dead load; the study of collapse of a star; the destruction of a crystal lattice; the description of self-organization and decay processes in biological systems; the simulation of turbulence; the analysis of chaotic movements in simple deterministic models; and many others. In all of the above listed examples, a slow change in system parameters, which is characterized by the presence of a small parameter in a system of differential equations, results in a change of the process quality. Sometimes such change may occur abruptly. To study the dynamics processes in such systems in their natural behavior, it is necessary to use adequate approaches to the qualitative analysis of solutions of the respective systems of equations with a small parameter

This chapter contains some results from the theory of differential and integral inequalities and the theory of stability of motion which form the basis for the approaches developed in the book and are intended for the study of dynamic properties of solutions of systems with a small parameter. From the numerous methods of nonlinear mechanics that have been developed recently, the methods of perturbations and averaging are actively used in this book (cf. Bogolyubov, Mitropolsky [1], Mitropolsky [1], Grebennikov [1], Volosov, Morgunov [1] and others).

Below readers will find more detailed bibliographic references that do not pretend to be exhaustive but provide an opportunity for an interested reader to approach the border beyond which a new area of research is awaiting.

1.2. Theorem 1.2.1 is a fundamental linear integral inequality known as the Gronwall or the Gronwall–Bellman inequality (see Bellman [1], Beesack [1]). In the process of formulating Theorems 1.2.2, 1.2.3, and 1.2.5 we follow the monograph by Lakshmikantham, Leela, Martynyuk [1]. Theorem 1.2.4 is stated as per Bihari [1]. Theorems 1.2.7 and 1.2.8 are available in the monograph by Martynyuk, Gutovsky [1]. Theorems 1.2.9–1.2.13 are taken from the monograph by Lakshmikantham , Leela, Martynyuk [1] (see Walter [1] and others).

1.3, 1.4. In the statement of the results of these sections, some results of the works of Krasovsky [1], Lyapunov [1], Lakshmikantham, Leela, Martynyuk [1], Rao [1], and Persidsky [1] were used. Many of the results related to the development of the comparison method can be found in the works of Conti [1], Corduneanu [2], Matrosov [2], and other authors.

1.5. The formulations of definitions of the μ-stability are given according to the article by Martynyuk [5] and the monograph by Martynyuk [16].

The works of Lagrange [1], Poincare [1], and Euler [1] were the basis for the creation and development of the current methods of the analysis of solutions of nonlinear systems with a small parameter.

Chapter 2

Analysis of the Boundedness of Motion

2.1 Introductory Remarks

The problem of the boundedness of motion of mechanical systems simulated by ordinary differential equations has been considered by many authors. Here we will only mention some works directly related to the investigation carried out in this chapter. In the work of Yoshizawa [1] the problem of the boundedness of solutions is considered in the context with the method of Lyapunov functions. In the work of Lakshmikantham and Leela [1] it was noted that the application of a perturbed Lyapunov function in the problem of the boundedness of motion results in the reduction of requirements to auxiliary functions in the study of nonuniform properties of motion. The extension of the conditions to perturbations of a Lyapunov function proposed in the article of Burton [1] provides an opportunity to consider the problem of the boundedness of motion of nonlinear weakly connected systems under wider assumptions on dynamical properties of subsystems.

Section 2.2 contains the statement of the problem and the main definitions of the μ-boundedness of motion with respect to two measures.

In Section 2.3, the general approach to the study of the μ-boundedness with respect to two measures is described. This approach is based on the application of the direct Lyapunov method and an auxiliary vector function.

In Section 2.4, the conditions for μ-boundedness are discussed, which were obtained by using the comparison technique. A scalar function constructed on the basis of the vector Lyapunov function is applied therein.

Section 2.5 contains the necessary and sufficient conditions for μ-boundedness of motion with respect to a part of variables of a weakly connected system.

Section 2.6 provides the criteria for different types of μ-boundedness of motion constructed via direct application of the vector Lyapunov function.

In Section 2.7, applications of certain results of this chapter are discussed. In particular, the Lienard oscillator, Lurie–Postnikov systems of connected equations, and nonlinear systems with weak linear connections between subsystems are discussed.

2.2   Statement of the Problem

A nonlinear system of ordinary differential equations that describes a weakly connected mechanical (or other) system with a finite number of degrees of freedom has the form

$$\frac{dx_s}{dt} = f_s(t, x_s, \mu g_s(t, x, \mu)),$$

$$x_s(t_0) = x_{s0}, \quad s = 1, 2, \ldots, m. \tag{2.2.1}$$

Here $x_s \in R^{n_s}$, $f_s \in C(R_+ \times R^{n_s} \times M \times R^{n_s}, R^{n_s})$, $g_s \in C(R_+ \times R^n \times M, R^{n_s})$, $n_1 + n_2 + \ldots + n_m = n$, $M = (0, 1]$, $\mu > 0$ is a small parameter.

In a particular case when the connections of subsystems are included additively, the system (2.2.1) is transformable to the following:

$$\frac{dx_s}{dt} = f_s(t, x_s) + \mu g_s(t, x_1, \ldots, x_m), \quad x_s(t_0) = x_{s0},$$

$$s = 1, 2, \ldots, m, \tag{2.2.2}$$

where $x_s \in R^{n_s}$, $f_s \in C(R_+ \times R^{n_s}, R^{n_s})$ and $g_s \in C(R_+ \times R^{n_1} \times \ldots \times R^{n_s}, R^{n_s})$.

If the vector functions f_s are linear at all $s = 1, 2, \ldots, m$, then the system (2.2.2) can be simplified yet more:

$$\frac{dx_s}{dt} = A_s(t)x_s + \mu g_s(t, x_1, \ldots, x_m), x_s(t_0) = x_{s0},$$

$$s = 1, 2, \ldots, m. \tag{2.2.3}$$

Here $A_s \in C(R_+, R^{n_s} \times R^{n_s})$, $A_s(t)$, $s = 1, 2, \ldots, m$, are matrices $(n_s \times n_s)$-continuous and bounded at all $t \in R_+$.

If the vector functions $g_s \equiv 0$ at all $s = 1, 2, \ldots, m$ or $\mu = 0$, then the system (2.2.2) consists of the independent subsystems

$$\frac{dx_s}{dt} = f_s(t, x_s), \quad x_s(t_0) = x_{s0},$$

$$s = 1, 2, \ldots, m. \tag{2.2.4}$$

To formulate definitions required for further analysis, we will characterize the dynamics of the k-th subsystem in the collection (2.2.2) by the two different measures $\rho_k(t, x_k)$ and $\rho_{k_0}(t, x_k)$, which take on values from the sets (cf. Movchan [1], Lakshmikantham and Salvadori [1])

$$M = \{\rho(t, x) \in C(R_+ \times R^n, R_+) : \inf_{t,x} \rho(t, x) = 0\},$$

$$M_0 = \{\rho(t, x) \in M : \inf_{x \in R^n} \rho(t, x) = 0 \quad \text{for any} \quad t \in R\}.$$

Here $\rho(t, x) = \sigma^T \tilde{\rho}(t, x)$, $\sigma^T = (\sigma_1, \ldots, \sigma_m)$ and the vector measure $\tilde{\rho}(t, x) = \rho_1(t, x_1), \ldots, \rho_m(t, x_m))^T$, $\sigma_s = \text{const} > 0$ at all $s = 1, 2, \ldots, m$.

Formulate the following definition.

Definition 2.2.1 The motion $x(t, \mu) = (x_1(t; t_0, x_0), \ldots, x_m(t; t_0, x_0))^{\mathrm{T}}$ of the system (2.2.2) is said to be:

(1) $(\rho_0, \rho)\mu$-equibounded, if at any values of $a \geq 0$ and $t_0 \in R_+$ there exists a positive function $\beta = \beta(t_0, a)$ continuous with respect to t_0 at all a and a value of $\mu^* = \mu^*(t_0, a) > 0$ such that

$$\rho(t, x(t, \mu)) < \beta \quad \text{at all} \quad t \geq t_0,$$

as soon as

$$\rho_0(t_0, x_0) \leq a \quad \text{and} \quad \mu < \mu^*(t_0, a);$$

(2) uniformly $(\rho_0, \rho)\mu$-bounded, if β and μ^* in definition (1) do not depend on t_0;

(3) uniformly ultimately $(\rho_0, \rho)\mu$-bounded, if it is uniformly $(\rho_0, \rho)\mu$-bounded and for any $a \geq 0$ and $t_0 \in R_+$ there exist positive numbers $\mu^* \in (0, 1]$, $\beta^* > 0$ and $\tau = \tau(t_0, a)$ such that $\rho(t, x(t, \mu)) < \beta^*$ at all $t \geq t_0 + \tau$, as soon as $\rho_0(t_0, x_0) \leq a$ and $\mu < \mu^*$.

Remark 2.2.1 Depending on the considered type of the boundedness of motion, the measures ρ_0 and ρ may be chosen by different methods. Here are some of the measures applied:

(1) in the study of the boundedness of motion in the sense of definitions from the monographs of Reissig et al. [1] and Yoshizawa [1], the measures $\rho_0(t, x) = \rho(t, x) = \|x\|$ are applied where $\| \cdot \|$ is the Euclidean norm of the vector x;

(2) in the study of the boundedness of the prescribed motion $x^*(t)$ the measures $\rho_0(t, x) = \rho(t, x) = \|x - x^*(t)\|$ are applied;

(3) in the study of the boundedness of motion with respect to a part of variables the measures $\rho_0(t, x) = \|x\|$ and $\rho(t, x) = \|x_k\|$, $1 \leq k \leq m$ are applied;

(4) in the study of the boundedness of motion with respect to a set A the measures $\rho_0(t, x) = \rho(t, x) = d(x, A)$ are applied, where $d(x, A)$ is the distance from the point x to the set $A \subset R^n$.

Thus, the conditions for the boundedness of motion of the system (2.2.2) under different assumptions on the dynamic properties of the subsystems (2.2.4) with respect to two different measures are generalized in relation to other conditions determined earlier, when measures (1)–(4) were used.

2.3 μ-Boundedness with Respect to Two Measures

In this section we will apply the functions of comparison of the classes K, L (see Definition 1.3.1) and:

$$CK = \{b \in C(R_+^2, R_+) \colon b(t, s) \in K \text{ at every } t\},$$
$$KL = \{\gamma \in C(R_+^2, R_+) \colon \gamma(t, s) \in K \text{ for every } s \text{ and}$$
$$\gamma(t, s) \in L \text{ for every } t\},$$
$$KR = \{c \in K \colon \lim_{u \to \infty} c(u) = \infty\}.$$

Note that the comparison functions from the above classes are widely used in works on the theory of stability of motion.

Below we will show some relations between the measures $\rho_0(t, x)$ and $\rho(t, x)$.

Definition 2.3.1 Let the measures ρ_{k_0}, $\rho_k \in M$ at all $k = 1, 2, \ldots, m$. We say that

(1) the measure $\rho(t, x) = \sum_{k=1}^{m} \rho_k(t, x_k)$ is continuous with respect to the measure $\rho_0(t, x) = \sum_{k=1}^{m} \rho_{k_0}(t, x_k)$, if there exists a constant $\Delta > 0$ and a comparison function $b \in CK$-class, such that $\rho(t, x) \leq b(t, \rho_0(t, x))$, as soon as $\rho_0(t, x) < \Delta$;

(2) the measure ρ is uniformly continuous with respect to the measure ρ_0, if in definition (1) the comparison function b does not depend on t;

(3) the measure ρ is asymptotically continuous with respect to the measure ρ_0, if there exists a constant $\Delta_1 > 0$ and a comparison function $\psi \in KL$-class, such that $\rho_0(t, x) \leq \psi(t, \rho_0(t, x))$, as soon as $\rho_0(t, x) < \Delta_1$.

For the independent subsystems (2.2.4) construct auxiliary functions $v_s(t, x_s)$, $s = 1, 2, \ldots, m$. Let $v_s \in C(R_+ \times R^{n_s}, R_+)$ at all $s = 1, 2, \ldots, m$. For any function $v_s \in C(R_+ \times R^{n_s}, R_+)$ determine the function

$$D^+ v_s(t, x_s) = \lim_{\theta \to 0^+} \sup \frac{1}{\theta} [v_s(t + \theta, x_s + \theta f_s(t, x_s)) - v_s(t, x_s)]$$

at all $s = 1, 2, \ldots, m$ for the values $(t, x_s) \in R_+ \times R^{n_s}$.

In order to note that the full derivative of the function $v_s(t, x_s)$ is calculated along the solutions of a certain subsystem (*), we will denote this as follows:

$$D^+ v_s(t, x_s)|_{(*)}, \quad s = 1, 2, \ldots, m.$$

Definition 2.3.2 A function

$$v(t, x, \mu) - \sum_{s=1}^{m} [v_s(t, x_s) + w_s(t, x_s, \mu)], \tag{2.3.1}$$

where $v \in C(R_+ \times R^n \times M, R_+)$, $v(t, x, \mu)$ is locally Lipshitz with respect to x, is said to be strengthened if $v(t, x, \mu)$ has a certain type of sign definiteness with respect to the measure ρ, while the functions $v_s(t, x_s)$ are constantly positive at all $s = 1, 2, \ldots, m$ and

$$|w_s(t, x_s, \mu)| < c_s(\mu), \ s = 1, 2, \ldots, m,$$

where $\lim\limits_{\mu \to 0} c_s(\mu) = 0$.

Definition 2.3.3 Let the function $v(t, x, \mu)$ be constructed by the formula (2.3.1). We say that the strengthened function $v(t, x, \mu)$:

(1) is ρ-positive definite, if there exist constants $\delta_1 > 0$, $\mu^* > 0$, and a comparison function $a \in K$-class, such that

$$a(\rho(t, x)) \le v(t, x, \mu),$$

as soon as $\rho(t, x) < \delta_1$ and $\mu < \mu^*$;

(2) is ρ-decrescent, if there exist constants $\delta_2 > 0$, $\tilde{\mu} > 0$, and a function $\omega \in K$-class such that

$$v(t, x, \mu) < \omega(\rho(t, x))$$

as soon as $\rho(t, x) < \delta_2$ and $\mu < \tilde{\mu}$.

Now pass on to the formulation and proof of statements on the boundedness of motion of the system (2.2.2) with respect to two different measures.

Theorem 2.3.1 *Assume that*:

(1) *for the independent subsystems (2.2.2) the measures $\rho_{0k}, \rho_k \in M$ are specified, and the measure $\rho(t, x)$ is continuous with respect to the measure $\rho_0(t, x)$;*

(2) *there exist constantly positive functions $v_k \in C(R_+ \times R^{n_k}, R_+)$ and functions $w_k \in C(R_+ \times R^{n_k} \times M, R_+)$ at all $k = 1, 2, \ldots, m$ such that the strengthened function $v(t, x, \mu)$ [see the formula (2.3.1)] is locally Lipshitz with respect to x and satisfies the estimates*

$$a(\rho(t, x)) \le v(t, x, \mu) \le r(t, \rho_0(t, x)) \tag{2.3.2}$$

at all $(t, x) \subset R_+ \times R^n$, where $a(\gamma) \to \infty$ at $\gamma \to \infty$ and $r \in C(R_+ \times R_+, R_+)$;

(3) *there exists a value of $\mu^* \in (0,1]$, at which the following inequality holds:*

$$D^+ v(t,x,\mu)|_{(2.2.2)} \leq 0 \quad \text{at all} \quad (t,x) \in R_+ \times R^n \quad \text{and at} \quad \mu < \mu^*.$$

Then the motion $x(t,\mu)$ of the weakly connected system (2.2.2) is $(\rho_0, \rho)\mu$-bounded.

Proof Let $a > 0$, $t_0 \in R_+$ be specified and let $x(t,\mu) = x(t; t_0, x_0, \mu)$ be the motion of the system (2.2.2) with its initial conditions satisfying the inequality

$$\rho_0(t_0, x_0) = \sum_{k=1}^{m} \rho_{0k}(t_0, x_{0k}) \leq a.$$

From condition (1) of Theorem 2.3.1 it follows that there exists a function b belonging to the CK-class, such that

$$\rho(t,x) \leq b(t, \rho_0(t,x)).$$

Choose $\beta = \beta(t_0, a) > 0$ so that

$$\beta > \max\{b(t_0, a), r^{-1}(t_0, a)\}. \tag{2.3.3}$$

From the estimate (2.3.2) at the chosen value of β, it is clear that $\rho(t_0, x_0) < \beta$. To prove the theorem, it suffices to show that

$$\rho(t, x(t,\mu)) < \beta \quad \text{at all} \quad t \geq t_0 \quad \text{and} \quad \mu < \mu^*. \tag{2.3.4}$$

Let the inequality (2.3.4) be satisfied not at all $t \geq t_0$. Then at a fixed μ^* there should exist $t_1 > t_0$ such that $\rho(t_1, x(t_1, \mu)) = \beta$ at $\mu < \mu^*$. Since the function $v(t,x,\mu)$ is nonincrescent, from conditions (2) and (3) of Theorem 2.3.1 it follows that

$$a(\beta) \leq v(t_1, x(t_1, \mu), \mu) \leq v(t_0, x_0, \mu) \leq r(t_0, a) \quad \text{at} \quad \mu < \mu^*.$$

This contradicts the choice of β by the formula (2.3.3). Consequently, the motion $x(t,\mu)$ of the weakly connected system (2.2.2) is $(\rho_0, \rho)\mu$-bounded.

Theorem 2.3.1 has a number of corollaries.

Corollary 2.3.1 Let the measures ρ_0 and ρ be defined as follows:

$$\rho_0(t, x_1, \ldots, x_m) = \sum_{k=1}^{m} \rho_{k0}(t, x_k),$$

$$\rho(t,x) = \tilde{\rho}(t, x_1, \ldots, x_l) = \sum_{k=1}^{l} \rho_k(t, x_k), \quad 1 \leq l < m,$$

and let all the conditions of Theorem 2.3.1 be satisfied.

Then the motion of the weakly connected system (2.2.2) is $(\rho_0, \tilde{\rho})\mu$-bounded, that is, bounded with respect to a part of variables $x_1, \ldots, x_l$.

Corollary 2.3.2 Let in the system (2.2.2) $\mu = 0$ and let conditions (2) and (3) of Theorem 2.3.1 be satisfied with the function

$$v(t, x, \mu) = v_0(t, x) = \sum_{k=1}^{m} v_k(t, x_k).$$

Then the motion of the independent subsystems (2.2.4) is $(\rho_0, \rho)\mu$-bounded.

Corollary 2.3.3 Let in the system (2.2.2) $\mu = 0$, $m = 1$ and let all the conditions of Theorem 2.3.1 be satisfied with the measures

$$\rho_0(t, x) = \rho_{10}(t, x_1) \in M,$$
$$\rho(t, x) = \rho_1(t, x_1) \in M$$

and the function

$$v(t, x, \mu) = v_1(t, x_1).$$

Then the motion of the system

$$\frac{dx_1}{dt} = f_1(t, x_1), \quad x_1(t_0) = x_{10} \tag{2.3.5}$$

is (ρ_{10}, ρ_1)-bounded.

Corollary 2.3.4 Let in the system (2.2.2) $\mu = 0$, $m = 1$, the function $r = 0$ and let all the conditions of Theorem 2.3.1 be satisfied with the measures

$$\rho_{10}(t, x_1) = \rho_1(t, x_1) = \|x_1\|$$

and the function

$$v(t, x, \mu) = v_1(t, x_1).$$

Then the motion of the system (2.3.5) is bounded.

Later we will consider the domains

$$S_k(\rho, \delta) = \{x_k \in R^{n_k} : \rho_k(t, x_k) < \delta\}, \quad k = 1, 2, \ldots, m,$$

and their contradomain $S_k^c(\rho, \delta)$. Let $S(\rho, \Delta) = \bigcup_{k=1}^{m} S_k(\rho, \delta)$ and $S^c(\rho, \Delta)$ be a contradomain of $S(\rho, \Delta)$.

Theorem 2.3.2 *Assume that:*

(1) *for the subsystems (2.2.2) the measures $\rho_{0k}, \rho_k \in M$ are specified, and the measure $\rho(t, x)$ is uniformly continuous with respect to the measure $\rho_0(t, x)$;*

(2) *there exist functions*

$$v_k \in C(S_k^c(\rho, \delta), R_+) \quad and \quad w_k \in C(S_k^c(\rho, \delta) \times M, R)$$

at all $k = 1, 2, \ldots, m$, a comparison function a from the K-class and a function $q \in C(R_+, R_+)$ such that the strengthened function $v(t, x, \mu)$ is locally Lipschitz with respect to x and

$$a(\rho(t, x)) \leq v(t, x, \mu) \leq q(\rho_0(t, x))$$

at all $(t, x) \in S^c(\rho, \Delta)$, where $a(\gamma) \to \infty$ at $\gamma \to \infty$;

(3) *there exists $\mu^* \in (0, 1]$, at which the following inequality holds*

$$D^+ v(t, x, \mu)|_{(2.2.2)} \leq 0 \quad at\ all \quad (t, x) \in S^c(\rho, \Delta) \quad and \quad \mu < \mu^*.$$

Then the motion $x(t, \mu)$ of the weakly connected system (2.2.2) is uniformly $(\rho_0, \rho)\mu$-bounded.

Proof From condition (1) of Theorem 2.3.2 it follows that there exists a function φ from the K-class such that

$$\varphi(t, x) \leq \varphi(\rho_0(t, x)). \tag{2.3.6}$$

For an arbitrary $a > 0$ choose $\beta = \beta(a) > 0$ so that

$$a(\beta) > \max\{q(a), q(\Delta), a^{-1}(\varphi(a)), a^{-1}(\varphi(\Delta))\}. \tag{2.3.7}$$

Now let $t_0 \in R_+$ and $\rho_0(t_0, x_0) < a$. Assume that for the motion $x(t, \mu) = (x_1(t; t_0, x_0, \mu), \ldots, x_m(t; t_0, x_0, \mu))^{\mathrm{T}}$ of the system (2.2.2) there exists t^* such that

$$\rho(t^*, x(t^*, \mu)) \geq \beta. \tag{2.3.8}$$

Then there exist values $t_1, t_2 \colon t_0 \leq t_1 \leq t_2 < t^*$, for which

$$\rho_0(t_1, x(t_1, \mu)) = \max\{a, \Delta\}, \quad \rho(t_2, x(t_2, \mu)) = \beta,$$
$$(t, x(t, \mu)) \in S(\rho, \beta) \cap S^c(\rho_0, \max\{a, \Delta\}), \quad t \in [t_1, t_2). \tag{2.3.9}$$

From condition (2) of Theorem 2.3.2 it follows that

$$v(t_1, x(t_1, \mu), \mu) \leq q(\rho_0(t_1, x(t_1, \mu))) = \max\{q(a), q(\Delta)\} \tag{2.3.10}$$

and

$$v(t_2, x(t_2, \mu), \mu) \geq a(\rho(t_2, x(t_2, \mu))) = a(\beta). \tag{2.3.11}$$

According to condition (3) of Theorem 2.3.2, there exists $\mu^* \in (0, 1]$ and the following estimate holds:

$$v(t_2, x(t_2, \mu), \mu) \leq v(t_1, x(t_1, \mu), \mu) \quad at \quad \mu < \mu^*. \tag{2.3.12}$$

Taking into account (2.3.10) and (2.3.11), from (2.3.12) obtain

$$a(\beta) \leq \max\{q(a), q(\Delta)\}. \tag{2.3.13}$$

This contradicts the selection of $a(\beta)$ by the formula (2.3.7) and proves Theorem 2.3.2.

Theorem 2.3.2 has some corollaries, too.

Corollary 2.3.5 Let the measures $\rho_0, \rho \in M$ and be chosen as specified in Corollary 2.3.1. If all the conditions of Theorem 2.3.2 are satisfied, the motion of the weakly connected system (2.3.2) is uniformly $(\rho_0, \tilde{\rho})\mu$-bounded.

Corollary 2.3.6 Let in the system (2.2.2) $\mu = 0$ and let conditions (2) and (3) of Theorem 2.3.2 be satisfied with the function $v(t, x, \mu)$ indicated in Corollary 2.3.2.
Then the motion of the independent subsystems (2.2.4) is uniformly (ρ_0, ρ)-bounded.

Corollary 2.3.7 Let in the system (2.2.2) $\mu = 0$, $m = 1$ and let the conditions of Theorem 2.3.2 be satisfied with the measures $\rho_{10}, \rho_1 \in M$ and the function $v(t, x, \mu) = v_1(t, x_1)$, indicated in Corollary 2.3.3.
Then the motion of the systems (2.3.14) is uniformly (ρ_{10}, ρ_1)-bounded.

Corollary 2.3.8 Let in the system (2.2.2) $\mu = 0$ and $m = 1$. If here all the conditions of Theorem 2.3.2 with the measures $\rho_{10}(t, x_1) = \rho_1(t, x_1) = \|x_1\|$ are satisfied, and the function $v(t, x, \mu) = v_1(t, x_1)$, then the motion of the system (2.2.14) is uniformly bounded.

Theorem 2.3.3 *Assume that:*

(1) *conditions (1) and (2) of Theorem 2.3.2 are satisfied;*

(2) *there exists $\mu^* \in (0, 1]$ and a comparison function c from the K-class such that*
$$D^+ v(t, x, \mu)|_{(2.2.2)} \leq -c(\rho_0(t, x))$$
at all $(t, x) \in S^c(\rho_0, \Delta)$ and $\mu < \mu^*$.

Then the motion of the weakly connected system (2.2.2) is uniformly ultimately $(\rho_0, \rho)\mu$-bounded.

Proof Under the conditions of Theorem 2.3.3 all the conditions of Theorem 2.3.2 are satisfied and therefore the motion of the system (2.2.2) is uniformly (ρ_0, ρ, μ)-bounded. This means that there exists $\beta^* > 0$ such that

$$\rho(t, x(t, \mu)) < \beta^* \quad \text{at all} \quad t \geq t_0,$$

as soon as $\rho_0(t_0, x_0) < \gamma$ and $\mu < \mu^*$.
Consider the motion $x(t, \mu)$ of the system (2.2.2) with the initial conditions

$\rho_0(t_0, x_0) < a$, where a is an arbitrary number such that $a > \gamma$. Then there exists a positive number $\beta = \beta(a) > 0$ such that

$$\rho(t, x(t, \mu)) < \beta \quad \text{at all} \quad t \geq t_0 \quad \text{and} \quad \mu < \mu_1.$$

Show that there exists $t^* \in [t_0, t_0 + \tau]$, where

$$\tau = \frac{q(a) + 1}{c(\gamma)},$$

such that $\rho_0(t^*, x(t^*, \mu)) < \gamma$ at $\mu < \mu_2$. If this is not correct, then

$$\rho_0(t, x(t, \mu)) \geq \gamma \quad \text{at all} \quad t \in [t_0, t_0 + \tau] \quad \text{and} \quad \mu < \mu_2.$$

In this case, condition (2) of Theorem 2.3.3 implies the estimate

$$v(t_0 + \tau, x(t_0 + \tau, \mu), \mu) \leq v(t_0, x_0, \mu) - c(\gamma)\tau,$$

which, together with the estimate

$$a(\rho(t, x)) \leq v(t, x, \mu) \leq q(\rho_0(t, x)) \quad \text{at all} \quad (t, x) \in S^c(\rho, \Delta)$$

results in the inequality

$$0 \leq q(a) - c(\gamma)\frac{q(a) + 1}{c(\gamma)} < 0.$$

The obtained contradiction proves that under the conditions $\rho_0(t_0, x_0) < a$ and $\mu < \mu^* = \min\{\mu_1, \mu_2\}$ the estimate $\rho(t, x(t, \mu)) < \beta^*$ holds at all $t \geq t_0 + \tau$ and $\mu < \mu^*$. The theorem is proved.

Like Theorems 2.3.1 and 2.3.2, Theorem 2.3.3 has a number of corollaries.

Corollary 2.3.9 Let the measures $\rho_0, \rho \in M$ be chosen as indicated in Corollary 2.3.1. If all the conditions of Theorem 2.3.3 are satisfied, then the motion of the weakly connected system (2.2.2) is uniformly ultimately $(\rho_0, \tilde{\rho})\mu$-bounded.

Corollary 2.3.10 Let in the system (2.2.2) $\mu = 0$ and let conditions (2) and (3) of Theorem 2.3.3 be satisfied with the function indicated in Corollary 2.3.2.

Then the motion of independent systems (2.2.4) is uniformly ultimately (ρ_0, ρ)-bounded.

Corollary 2.3.11 Let in the system (2.2.2) $\mu = 0$, $m = 1$ and all the conditions of Theorem 2.3.3 be satisfied with the measures $\rho_{10}, \rho_1 \in M$ and the function $v(t, x, \mu) = v_1(t, x_1)$, indicated in Corollary 2.3.3. Then the motion of the system (2.2.14) is uniformly ultimately (ρ_{10}, ρ_1)-bounded.

Corollary 2.3.12 Let in the system (2.2.2) $\mu = 0$ and $m = 1$. If all the conditions of Theorem 2.3.3 with the measures $\rho_{10}(t, x_1) = \rho_1(t, x_1) = \|x_1\|$

and the function $v(t, x, \mu) = v_1(t, x_1)$ are satisfied, then the motion of the system (2.2.14) is uniformly ultimately bounded.

The general theorems 2.3.1–2.3.3 on the boundedness of motion with respect to two different measures ρ_0, ρ may provide the basis for the construction of different sufficient conditions for the boundedness of motion of the nonlinear weakly connected systems (2.2.2). At specifically chosen measures $\rho_0, \rho \in M$ and functions v_k and w_k at $k = 1, 2, \ldots, m$ the sufficient conditions for the boundedness of motion coincide in particular cases with those obtained both for second-order systems (see Reissig et al. [1]) and for a system of n ordinary differential equations (see Yoshizawa [2]).

2.4 Boundedness and the Comparison Technique

The comparison technique allows us to simplify the investigation of the boundedness of motion of the weakly connected system (2.2.2) by substituting it with the analysis of solutions of a nonlinear scalar comparison equation. This fruitful approach is based on theorems on differential inequalities (see Chapter 1).

2.4.1 Auxiliary results

Consider the scalar differential equation

$$\frac{du}{dt} = g(t, u, \mu), \quad u(t_0) = u_0 \geq 0. \tag{2.4.1}$$

Here $g \in C(R_+ \times R \times M, R)$, $g(t, u, \mu) = 0$ at all $t \geq t_0$, if and only if $u = 0$.

Definition 2.4.1 Let $\gamma(t, \mu)$ be a solution of the comparison equation (2.4.1), existing on the interval $J = [t_0, t_0 + a)$, $0 < a < +\infty$, $\mu \in M$. The solution $\gamma(t, \mu)$ is said to be the μ-maximum solution for the equation (2.4.1), if for any other solution $u(t, \mu) = u(t; t_0, u_0, \mu)$ of the equation (2.4.1), existing on J, the following inequality holds:

$$u(t, \mu) \leq \gamma(t, \mu) \quad \text{at all} \quad t \in J, \quad \mu \leq \mu_0. \tag{2.4.2}$$

The μ-minimum solution is obtained in a similar way, the sign in the inequality (2.4.2) is substituted with the opposite one.

Lemma 2.4.1 *Let the function $g \in (R_+ \times R \times M, R)$ and let $\gamma(t, \mu) = \gamma(t, t_0, u_0, \mu)$ be the maximum solution of the equation (2.4.1), defined on the interval J. Assume that the function $m \in C(R_+, R_+)$ and*

$$Dm(t) \leq g(t, m(t), \mu), \quad t \in J, \tag{2.4.3}$$

where D is a fixed Dini derivative.
Then

$$m(t) \leq \gamma(t, \mu), \quad t \in J,$$

as soon as $m(t_0) \leq u_0$.

Lemma 2.4.2 *Let the function* $g \in C(R_+ \times R \times M, R)$ *and let* $r(t, \mu) = r(t; t_0, u_0, \mu)$ *be the minimum solution of the equation (2.4.1), defined on J. Assume that the function* $n \in C(R_+, R_+)$ *and*

$$Dn(t) \geq g(t, n(t), \mu), \quad t \in J.$$

Then

$$n(t) \geq r(t, \mu), \quad t \in J,$$

as soon as $n(t_0) \geq u_0$.

Lemma 2.4.3 *Let the function* $v \in C(R_+ \times R^n, R_+)$ *and, in addition,* $v(t, x)$ *let it be locally Lipschitz with respect to x at every* $t \in R_+$. *Assume that the function* $D^+ v(t, x)$ *satisfies the inequality*

$$D^+ v(t, x) \leq g(t, v(t, x), \mu), \quad (t, x) \in R_+ \times R^n, \qquad (2.4.4)$$

where $g \in C(R_+ \times R_+ \times M, R)$.

Let $\gamma(t, \mu) = \gamma(t; t_0, u_0, \mu)$ *be the maximum solution of the equation (2.4.1) existing on* J_1.

Then for any solution $x(t) = x(t, t_0, x_0)$ *of the system*

$$\frac{dx}{dt} = f(t, x), \quad x(t_0) = x_0, \qquad (2.4.5)$$

existing on J_2, *the following estimate holds*

$$v(t, x(t)) \leq \gamma(t, \mu) \quad \text{at all} \quad t \in J_1 \cap J_2, \quad \mu \leq \mu_0, \qquad (2.4.6)$$

as soon as

$$v(t_0, x_0) \leq u_0.$$

The proofs of Lemmas 2.4.1–2.4.3 are given in the monograph by Lakshmikantham, Leela, and Martynyuk [1].

2.4.2   Conditions for the boundedness of motion

We will set out one variant of sufficient conditions for the boundedness of motion of the weakly connected system (2.2.2), which is based on the comparison technique.

The motion of the system (2.2.2) will be considered in the space $R^n = R^{n_1} \times R^{n_1} \times \ldots \times R^{n_m}$. Let $E \subset R^n$, and let the sets $\bar{E}$, E^c, and ∂E be the

closure, the complement, and the boundary of E. For an arbitrary $H > 0$, define an open ball

$$S(H) = \{x \in R^n : \|x\| < H\},$$

where $\|\cdot\|$ is the Euclidean norm of the state vector $x(t)$ of the system (2.2.2). Introduce the following assumptions:

(1) the dynamics of the subsystem (2.2.2) are characterized by the functions $v_s \in C(R_+ \times E^c, R_+)$, the functions $v_s \geq 0$ at all $s = 1, 2, \ldots, m$ are locally Lipschitz with respect to x_s;

(2) the estimate of the influence of the connection functions $\mu g_s(t, x)$, $s = 1, 2, \ldots, m$, in the system (2.2.2) is taken into account in the functions $w_s(t, x, \mu)$ which are defined at all $(t, x, \mu) \in R_+ \times S^c \times M$;

(3) the functions

$$v_0(t, x) = \eta^T V(t, x), \quad \eta \in R_+^m,$$

where $V(t, x) = (v_1(t, x_1), \ldots, v_m(t, x_m))^T$, and

$$w_0(t, x, \mu) = \eta^T W(t, x, \mu),$$

where $W(t, x, \mu) = (w_1(t, x, \mu), \ldots, w_m(t, x, \mu))^T$, satisfy special conditions.

Theorem 2.4.1 *Assume that:*

(1) *the set $E \subset R^n$ is compact and for the subsystem (2.2.2) there exist functions $v_s(t, x_s) \geq 0$, $s = 1, 2, \ldots, m$, such that the function $v_0(t, x)$ is locally Lipschitz with respect to x at every $t \in R_+$, the comparison functions $a, b \in K$-class, $a(r) \to \infty$ at $r \to \infty$, and the function $\Phi \in C(R_+ \times R_+, R)$ are such that*

(a) *$a(\|x\|) \leq v_0(t, x) \leq b(\|x\|)$ at all $(t, x) \in R_+ \times E^c$,*

(b) *$D^+ v_0(t, x)|_{(2.2.2)} \leq \Phi(t, v_0(t, x), \mu)$ at all $(t, x) \in R_+ \times E^c$, where $\Phi(t, u, \mu) = 0$ at $u = 0$;*

(2) *there exist functions $w_s(t, x, \mu)$, $s = 1, 2, \ldots, m$, such that the function $w_0(t, x, \mu)$ is locally Lipschitz with respect to x at every $t \in R_+$ and the following estimates hold:*

(a) *$|w_0(t, x, \mu)| \leq c(\mu)$ at $(t, x, \mu) \in R_+ \times S^c(H) \times M$;*

(b) *$D^+ v_0(t, x)|_{(2.2.2)} + D^+ w_0(t, x, \mu)|_{(2.2.2)} \leq \Psi(t, v_0(t, x) + w_0(t, x, \mu), \mu)$, where $c(\mu)$ is a nondecrescent function μ, $\lim_{\mu \to 0} c(\mu) = 0$ and $\Psi \in C(R_+ \times R \times M, R)$;*

(3) *the solution of the comparison equation*

$$\frac{du}{dt} = \Phi(t, u, \mu), \quad u(t_0) = u_0 \geq 0, \tag{2.4.7}$$

is μ-bounded;

(4) *the solution of the comparison equation*

$$\frac{dw}{dt} = \Psi(t, w, \mu), \quad w(t_0) = w_0 \geq 0, \tag{2.4.8}$$

is uniformly μ-bounded.

Then the motion $x(t, \mu)$ of the system (2.2.2) is μ-bounded.

Proof In view of the fact that E is compact, there exists $H_0 > 0$ such that $S(H_0) \supset S(E, H)$ for some $H > 0$. Here $S(E, H) = \{x \in R^n : d(x, E) < H\}$, $d(x, E) = \inf_{y \in E} \|x - y\|$. Assume that $t_0 \in R_+$ and $a \geq H_0$. Let

$$a_1 = a_1(t_0, a) = \max\{a_0, a^*\}, \tag{2.4.9}$$

where $a_0 = \max(v_0(t_0, x_0) : x_0 \in \overline{S(a)} \cap E^c)$ and $a^* \geq v_0(t, x)$ at $(t, x) \in R_+ \times \partial E$.

Since all solutions of the equation (2.4.7) are μ-bounded, for a specified $t_0 \in R_+$ and $a_1 > 0$ there exists $\beta_0 = \beta_0(t_0, a_1)$ and $\mu_1 = \mu_1(a_1)$ such that

$$u(t, t_0, u_0, \mu) < \beta_0 \quad \text{at all} \quad t \geq t_0, \tag{2.4.10}$$

as soon as $u_0 < a_1$ and $\mu < \mu_1$. From condition (4) of Theorem 2.4.1 it follows that at a specified $a_2 > 0$ there exists $\beta_1(a_2)$ and $\mu_2(a_2)$ such that

$$w(t; t_0, w_0, \mu) < \beta_1(a_2) \quad \text{at all} \quad t \geq t_0, \tag{2.4.11}$$

as soon as $w_0 < a_2$ and $\mu < \mu_2$. Let $u_0 = v_0(t_0, x_0)$ and $a_2 = b(a) + \beta_0$. Since $a(r) \to \infty$ at $r \to \infty$, it is possible to choose $\beta = \beta(t_0, a)$ so that

$$a(\beta) > \beta_1(a). \tag{2.4.12}$$

Now show that if $x_0 \in S(a)$, then the solution $x(t, \mu)$ of the system (2.2.2) satisfies the inclusion $x(t, \mu) \in S(\beta)$ at all $t \geq t_0$ and $\mu < \mu_0$. If this is not correct, then one can find a solution $x(t, \mu)$ such that for some $t^* \geq t_0$ at $\mu < \mu^0$ the relation $\|x(t^*, \mu)\| = \beta$ would hold. Since $S(E, H) \subset S(H_0)$, it is necessary to consider two cases:

(a) the inclusion $x(t, \mu) \in E^c$ is true at all $t \in [t_0, t^*]$ and $\mu < \mu^0$;

(b) there exists $\tilde{t} \geq t_0$ such that $x(\tilde{t}, \mu) \in \partial E$ and $x(t, \mu) \in E^c$ at $t \in [\tilde{t}, t^*]$ and $\mu < \mu_0$.

First, consider case (a). For the inclusion (a) one can find $t_1 > t_0$ such that

$$\begin{cases} x(t_1, \mu) \in \partial S(H), \\ x(t*, \mu) \in \partial S(\beta), \\ x(t, \mu) \in S^0(H) \quad \text{at} \quad t \in [t_1, t*], \mu < \mu_0. \end{cases} \quad (2.4.13)$$

Denoting $m(t, \mu) = v(t, x) + |w(t, x, \mu)|$, where $m \in C(R_+ \times M, R_+)$, obtain the inequality

$$D^+ m(t, \mu) \leq \Psi(t, m(t, \mu), \mu), \quad t \in [t_0, t*] \quad (2.4.14)$$

and the estimate

$$m(t, \mu) \leq w^+(t; t_1, u_0, \mu), \quad t \in [t_0, t*], \quad (2.4.15)$$

where $w^+(t_1; t_1, u_0, \mu) = u_0$, w^+ is the maximum solution of the equation (2.4.8). Thus,

$$m(t^*, \mu) \leq w^+(t^*; t_1, m(t_1, \mu)). \quad (2.4.16)$$

Similarly, using the inequality from condition (1b) of Theorem 2.4.1 and the comparison equation (2.4.7), obtain

$$(t_1, x(t_1, \mu)) \leq u^+(t_1; t_0, v(t_0, x(t_0, \mu))) \quad (2.4.17)$$

at all $t \in [t_0, t_1]$, where u^* is the maximum solution of the equation (2.4.7). It is obvious that if we choose

$$u_0 = v_0(t_0, x_0) < a_1,$$

then, in compliance with (2.4.10), we will obtain

$$u^+(t_1; t_0, v_0(t_0, x_0)) \leq \beta_0. \quad (2.4.18)$$

Now take a $\mu_3 \in M$ from the formula $\mu_3 = c^{-1}(\beta_0)$. Then, in view of condition (2a) of Theorem 2.4.1, obtain

$$|w_0(t, x, \mu)| \leq c(\mu) = c(c^{-1}(\beta_0)) = \beta_0. \quad (2.4.19)$$

Hence

$$w_0 = v_0(t_1, x(t_1; t_0, x_0)) + |w_0(t_1, x(t_1; t_0, x_0), \mu)| < b(a) + \beta_0 = a_2$$
$$\text{at} \quad \mu < \mu_3.$$

From the estimate (2.4.15) obtain

$$v_0(t_1, x(t_1; t_0, x_0)) + |w_0(t_1, x(t_1; t_0, x_0), \mu)|$$
$$< w^+(t_1, t_0, w_0, \mu) < \beta_1(a_2) \quad \text{at} \quad \mu < \mu_3.$$

Hence, taking into account (2.4.12), it follows that

$$a(\beta) + \beta_0 < \beta_1(a_2) < a(\beta) \quad \text{at} \quad \mu < \mu_3. \quad (2.4.20)$$

The obtained contradiction (2.4.20) shows that $x(t, \mu) \in S(\beta)$ at all $t \geq t_0$ and $\mu < \mu^0 = \min(\mu_1, \mu_2, \mu_3)$.

Let case (b) be realized. Here, like before, we obtain the inequality (2.4.15), where $t_1 \geq t_0$ satisfies the inclusion (2.4.13). For the function $v_0(t, x)$ the following estimate holds:

$$v_0(t_1, x(t_1; t_0, x_0)) < u^+(t; \tilde{t}, v_0(\tilde{t}, x(\tilde{t}; t_0, x_0))).$$

In case (b) we have $x(\tilde{t}; t_0, x_0, \mu) \in \partial E$ and $v_0(\tilde{t}, x(\tilde{t}; t_0, x_0, \mu)) \leq a^* < a_1$. Therefore, the reasoning is similar to the above results in the contradiction (2.4.20). Hence it follows that if $x_0 \in S(a)$ and $a \geq H$, $\mu < \mu^0$, then $\|x(t; t_0, x_0, \mu)\| < \beta$ at all $t \geq t_0$. At $a < H$ we assume that $\beta(t_0, a) = \beta_0(t_0, H)$.

Theorem 2.4.1 is proved.

Condition (1a) from Theorem 2.4.1 is essential for the proof of the μ-boundedness of motion of the system (2.2.2). If $a(r)$ does not tend to $+\infty$ at $r \to +\infty$, then the function $v_0(t, x)$ is not radially unbounded, and therefore its application in the study of the μ-boundedness of the system (2.2.2) is impossible.

Show the method of application of a function $v_0(t, x)$ that does not satisfy property (1a).

Definition 2.4.2 (cf. Burton [1]). The function $v_0(t, x)$, $v_0 \colon R_+ \times R^n \to R_+$ is strengthened by the function $u \colon R^n \times M \to R_+$, if the function

$$v_0(t, x) + u(x, \mu) \tag{2.4.21}$$

is radially unbounded, and the following conditions are satisfied:

(1) there exist disjoint open sets $S_1, \ldots, S_k$ in R^n and continuous functions $u_1, \ldots, u_k$, $u_i \colon S_i \times M \to R_+$ which have continuous partial derivatives in S_i, and

$$u(x, \mu) = \begin{cases} u_i(x, \mu), & \text{if } x \in S_i \text{ for some } i, \\ 0, & \text{if } x \in (\cup S_i)^c; \end{cases}$$

(2) there exist positive constants $L_1, \ldots, L_k$ such that for every i at $0 < L_i^* < L_i$ there exists such a constant $D > 0$ that if $x \in S_i$ and $v_0(t, x) \leq L_{*i}$, then $u_i(x, \mu) \leq D$.

Note that if the function (2.4.2) is radially unbounded for every $L > 0$, then there exists a constant $H > 0$ such that if

$$v_0(t, x) \leq L \quad \text{and} \quad \|x\| \geq h, \tag{2.4.22}$$

then $x \in S_i$ for some i.

Like in Theorem 2.4.1, the function $v_0(t, x)$ is determined by the formula

$$v_0(t, x) = \sum_{s=1}^{m} a_s v_s(t, x_s),$$

and the functions $v_i(x, \mu)$ and hence the function $u(x, \mu)$ are constructed with consideration for the connection functions $\mu g_s(t, x_1, \ldots, x_m)$ between the subsystems.

Theorem 2.4.2 *Assume that the motion equations (2.2.2) are such that:*

(1) *for the subsystems (2.2.4) there exist auxiliary functions $v_s(t, x_s)$ such that for the function $v_0(t, x)$ the following conditions are satisfied:*

    (a) $v_s\colon R_+ \times R^{n_s} \to R_+$, v_s *have continuous partial derivatives on* $R_+ \times R^{n_s}$,

    (b) *there exists a nonnegative constant M such that in the domain* $R_+ \times S^c(M)$ *the following inequality holds:*

$$\left. \frac{dv_0(t, x)}{dt} \right|_{(2.2.2)} \leq 0,$$

    (c) *if $M > 0$, then there exist positive constants K and P, $P > M$, such that at all $t \geq 0$*

$$v_0(t, x) \leq K \quad at \quad \|x\| = M$$

    and

$$v_0(t, x) > K \quad at \quad \|x\| = P;$$

(2) *there exists a function $u(x, \mu)$, which strengthens the function $v_0(t, x)$, and for every $x \in \bigcup_i S_i$ there exists $\mu^0 \in M$ such that the inequality*

$$\left. \frac{du(x, \mu)}{dt} \right|_{(2.2.2)} \leq 0$$

holds at all $t \geq 0$ and $\mu < \mu^0$.

Then the motion $x(t, \mu)$ of the system (2.2.2) is μ-bounded.

Proof If $x \in \bigcup S_i$, then for some i the motion $x \in S_i$. By assumption the sets S_i are open so that there exists $\mathrm{grad}\, u(x, \mu) = \mathrm{grad}\, u_i(x, \mu)$.

If Theorem 2.4.2 is incorrect, then there exists a motion of the system (2.2.2) for which the vector function $x(t, \mu) = (x_1(t, \mu), \ldots, x_m(t, \mu))^{\mathrm{T}}$, defined on the maximum right-hand interval $[t_0, T)$, at $\mu < \mu^0$, is μ-unbounded. In this case there exists an increscent sequence $\{\tau_n\}\colon t_0 < \tau_n < T$ such that $\|x(\tau_n)\| \to \infty$ at $n \to \infty$.

From conditions (1b) and (1c) of Theorem 2.4.2 it follows that $\|x(t,\mu)\| \geq M$, and the condition

$$\left. \frac{dv_0(t,x)}{dt} \right|_{(2.2.2)} \leq 0$$

results in the estimate

$$v_0(t, x(t,\mu)) \leq v_0(t_0, x(t_0,\mu)) = L.$$

This means that there exists $H > 0$ such that the condition $\|x(t,\mu)\| \geq H$ implies $x(t,\mu) \in S_i$ for some i. Consequently, for all sufficiently large n a solution $x(\tau_n, \mu)$ belongs to some S_i. From the sequence $\{\tau_n\}$ one can choose a subsequence $\{\tau_n^*\}$ so that $x(\tau_n^*, \mu) \in S_i$ for some fixed i and $\|x(\tau_n^*, \mu)\| > H$. As soon as $v_0(\tau_n^*, x(\tau_n^*, \mu)) \leq L$ and the function $v_0(t, x) + u(x, \mu)$ is radially unbounded, the fact that $\|x(\tau_n^*, \mu)\| \to \infty$, means that $u(x(\tau_n^*, \mu), \mu)' = u_i(x(\tau_n^*, \mu), \mu) \to \infty$. Here we should consider the following two cases:

(a) there exists $t_1 \in [t_0, T)$ such that

$$x(t,\mu) \in S_i \quad \text{on} \quad [t_1, T);$$

(b) there exists a sequence $\{t_n\}$: $t_n < \tau_n < t_{n+1}$ such that

$$x(t,\mu) \in S_i \quad \text{on} \quad (t_n, t_{n+1}) \quad \text{and} \quad \|x(t_n,\mu)\| = H.$$

For the verification of the above two cases, note that the functions $u_i(x,\mu)$ are continuous on $\bar{S}_i$, $\|x(\tau_n)\| > H$ and the motion $x(t,\mu)$ is continuous. Since S_i are open and do not overlap in view of the condition (2.4.22), there exists a sequence $\{t_n\}$ such that $\|x(t_n,\mu)\| = H$ and $\|x(t,\mu)\| > H$ on $[t_n, \tau_n]$. The continuous function $x(t,\mu)$ cannot leave S_i at $t < \tau_n$, if only $\|x(t,\mu)\|$ does not reach H at some point in time $t = t_n$.

Let case (a) be true. Then

$$\left. \frac{du_i(x,\mu)}{dt} \right|_{(2.2.2)} \leq 0$$

and hence

$$u_i(x(t,\mu),\mu) \leq u_i(x(t_1,\mu),\mu)$$

on the interval $[t_1, T)$, which contradicts the possible unboundedness of the function $u_i(x(t,\mu),\mu)$.

Let the case (b) be true. Then $\|x(t_n,\mu)\| = H$ and there exists a subsequence $\{x(t_{n_k},\mu)\}$ with some limit $y \in \bar{S}_i$. Hence $u_i(x(t_n,\mu),\mu) \to u_i(y,\mu)$ at $t_n \to +\infty$. The condition

$$\left. \frac{du_i(x,\mu)}{dt} \right|_{(2.2.2)} \leq 0$$

on $[t_{n_k}, \tau_{n_k}]$ for sufficiently large k results in the estimate

$$u_i(x(\tau_n, \mu), \mu) \leq u_i(y, \mu) + 1,$$

which in its turn contradicts the possible unboundedness of the functions $u_i(x(\tau_n, \mu), \mu)$.

The theorem is proved.

Definition 2.4.3 The function $v(t, x, \mu)$, $v: R_+ \times R^n \times M \to R_+$, is a strengthened Lyapunov function if $v_0(t, x)$ is strengthened by the function $u(x, \mu)$ indicated in Definition 2.4.2, and

(1) for the function $v_0(t, x) = \sum_{s=1}^{m} a_s v_s(t, x)$ the following conditions are satisfied:

   (a) $v_s: R_+ \times R^{n_s} \to R_+$, v_s have continuous first-order partial derivatives,

   (b) there exists a nonnegative constant M such that in the domain $R_+ \times S^c(M)$ the following inequality holds

   $$\frac{dv_0(t, x)}{dt}\bigg|_{(2.2.2)} \leq 0,$$

   (c) if $M > 0$, then there exist positive constants K and P $(P > M)$ such that at all $t \geq 0$ the following inequalities hold:

   $$v_0(t, x) \leq K \quad \text{at} \quad \|x\| = M$$

   and

   $$v_0(t, x) > K \quad \text{at} \quad \|x\| = P;$$

(2) for every i and every $L > L_i$ there exists a positive constant J and continuous functions $\Phi: (0, L - L_i) \to R_+$ and $H: [J, \infty) \to R_+$, for which

$$\int_{0+}^{L-L_i} \frac{ds}{\Phi(s)} < \infty \quad \text{and} \quad \int_{J}^{\infty} H(s)\, ds = \infty, \qquad (2.4.23)$$

while the conditions $u_i(x, \mu) \geq J$ and $L \geq v(t, x, \mu) > L_i$ imply

$$\frac{dv_0(t, x, \mu)}{dt}\bigg|_{(2.2.2)} \leq -\Phi(v(t, x, \mu) - L_i)H(u_i(x, \mu)) \qquad (2.4.24)$$
$$\times \|\operatorname{grad} u_i(x, \mu)(f(t, x) + \mu g(t, x_1, \ldots, x_m))\|.$$

Now consider the following statement.

Theorem 2.4.3 *Let the motion equations (2.2.2) be such that:*

(1) *there exists a strengthening function $u(x, \mu)$ such that $v(t, x, \mu)$ is a strengthened Lyapunov function in accordance with Definition 2.4.3;*

(2) *for every i there exists $\mu^0 \in M$ such that if $u_i(x, \mu) \geq J$ and $v_0(t, x, \mu) = L_i$, then*

$$\frac{dv_0(t, x, \mu)}{dt}\Big|_{(2.2.2)} < 0 \quad at \quad \mu < \mu^0.$$

Then the motion $x(t, \mu)$ of the system (2.1.2) is μ-bounded.

The proof of this theorem is similar to that of Theorem 2.4.2.

2.5   Boundedness with Respect to a Part of Variables

Continue the study of the μ-boundedness of motion of the system (2.2.2) under the following assumptions:

(a) the right-hand parts of the system (2.2.2) are continuous and satisfy the conditions for the existence of the unique solution $x(t, \mu) = x(t; t_0, x_0, \mu)$ in the domain

$$t \geq 0, \quad \|x\| = \sum_{s=1}^{m} \|x_s\| < +\infty, \quad \mu \in M,$$

here $f_s(t, 0) \neq 0$ and $g_s(t, 0, \dots, 0) \neq 0$ for at least one $s = 1, 2, \dots, m$;

(b) any solution $x(t; t_0, x_0, \mu)$ of the system (2.2.2) is defined at all $t \geq t_0$ and $\mu \in M^0 \subset M$.

Taking into account Definition 2.2.1 and condition (3) from Remark 2.2.1, formulate some definitions of μ-boundedness of motion of the system (2.2.2) with respect to variables of a part of subsystems.

Represent the vector $x = (x_1^\mathrm{T}, \dots, x_m^\mathrm{T})^\mathrm{T}$ with subvectors x_s, $s = 1, 2, \dots, m$, as follows:

$$x = (y^\mathrm{T}, z^\mathrm{T})^\mathrm{T},$$

where

$$y^\mathrm{T} = (x_1^\mathrm{T}, \dots, x_k^\mathrm{T})^\mathrm{T} \quad and \quad z^\mathrm{T} = (x_{k+1}^\mathrm{T}, \dots, x_m^\mathrm{T})^\mathrm{T}.$$

Now we will give the following definitions.

Definition 2.5.1 The motion

$$x(t, \mu) = (x_1^\mathrm{T}(t; t_0, x_0, \mu), \dots, x_m^\mathrm{T}(t; t_0, x_0, \mu))^\mathrm{T}$$

of the system (2.2.2) is said to be:

(1) μ-bounded with respect to the subvector of variables $y = (x_1^T, \ldots, x_k^T)^T$, if for any $t_0 \geq 0$ and $x_0 = (x_{10}^T, \ldots, x_{m0}^T)^T$ one can find $N(t_0, x_0) > 0$ and $\mu_0 \in M$ such that

$$\|y(t; t_0, x_0, \mu)\| \leq N \quad \text{at all} \quad t \geq t_0 \quad \text{and} \quad \mu < \mu^0; \qquad (2.5.1)$$

(2) μ-bounded uniformly with respect to t_0 with respect to the subvector of variables $y = (x_1^T, \ldots, x_k^T)^T$, if in Definition 2.5.1 (1) for any x_0 one can choose $N(x_0) > 0$ independent of t_0;

(3) μ-bounded uniformly with respect to x_0 with respect to the subvector of variables $y = (x_1^T, \ldots, x_k^T)^T$, if for any $t_0 \geq 0$ and a compact set $E \subset R^n$ one can find $N(t_0, E) > 0$ and $\mu^0 \in M$ such that $x_0 \in E$ would imply the estimate (2.5.1);

(4) μ-bounded uniformly with respect to (t_0, x_0), if in Definition 2.5.1 (3) for any compact set E one can choose $N(E) > 0$ independent of t_0.

Now for the system (2.2.2) consider the function (2.4.21), that is, the function $v_0(t, x)$ strengthened by the function $u(x, \mu)$.

Theorem 2.5.1 *Assume that the motion equations (2.2.2) are such that:*

(1) *the strengthened function*

$$v_0(t, x) + u(x, \mu), \quad \mu < \mu^* \in M, \qquad (2.5.2)$$

in the range of values $(t, x) \in R_+ \times R^n$ satisfies the condition

$$a(\|y\|) \leq v_0(t, x) + u(x, \mu), \quad a(r) \to \infty \quad \text{at} \quad r \to \infty; \qquad (2.5.3)$$

(2) *there exists $\mu^* \in M$ such that for any motion $x(t; t_0, x_0, \mu)$ the function $v_0(t, x(t; t_0, x_0, \mu)) + u(x(t; t_0, x_0, \mu), \mu)$ is not increscent at all $t \geq t_0$ and at $\mu < \mu^*$.*

Then the motion $x(t, \mu)$ of the system is μ-bounded with respect to the subvector of variables $y = (x_1^T, \ldots, x_k^T)^T$.

Proof. Sufficiency According to condition (1) of Theorem 2.5.1, at any $t \geq t_0$ and $x_0 \in R^n$ for the number $\eta = v_0(t_0, x_0) + u(x_0, \mu)$ at $\mu < \mu^*$ one can choose $N(\eta) = N(t_0, x_0) > 0$ so that if $\|y\| > N$, then $a(\|y\|) \leq v_0(t_0, x_0) + u(x_0, \mu)$ at $\mu < \mu^*$.

Condition (2) of Theorem 2.5.1 implies that at $\mu < \mu^*$

$$a(\|y(t, \mu)\|) \leq v_0(t, x(t; t_0, x_0, \mu)) + u(x(t; t_0, x_0, \mu), \mu) \leq \eta \leq a(N).$$

Hence $\|y(t, \mu)\| < N$ at all $t \geq t_0$ and $\mu < \mu^*$.

Necessity From the fact that the motion $x(t,\mu)$ of the system (2.2.2) is μ-bounded with respect to the subvector $y = (x_1^T,\ldots,x_k^T)^T$ it follows that in the domain $R_+ \times R^n$ for the function $v_0(t,x) + u(x,\mu)$ there exists

$$\sup_{\tau \geq 0} \|y(t+\tau;t,x,\mu)\| = v_0(t,x) + u(x,\mu). \tag{2.5.4}$$

It is clear that there exists $\mu^* \in M$ at which $v_0(t,x)+u(x,\mu) \geq \|y\|$, if $\mu < \mu^*$. For the values $t_1 < t_2$ obtain

$$v_0(t_1, x(t_1;t_0,x_0,\mu)) + u(x(t_1;t_0,x_0,\mu),\mu)$$
$$= \sup_{\tau \geq 0} \|y(t_1+\tau;t,x,\mu)\| \geq \sup_{\tau \geq 0} \|y(t_2+\tau;t_0,x_0)\|$$
$$= v_0(t, x_0(t_2;t_0,x_0,\mu)) + u(x(t_2;t_0,x_0,\mu),\mu)).$$

This means that the function $v_0(t, x(t;t_0,x_0,\mu)) + u(x(t;t_0,x_0,\mu),\mu)$ is not increscent.

Theorem 2.5.1 is proved.

Theorem 2.5.2 *Assume that the motion equations (2.2.2) are such that:*

(1) *the strengthened function $v_0(t,x)+u(x,\mu)$ in the range of values $(t,x) \in R_+ \times R^n$ satisfies the condition (2.5.3) and, in addition,*

$$v_0(t,x) + u(x,\mu) \leq w(x) \quad at \quad \mu < \mu^*, \tag{2.5.5}$$

where $w(x)$ is a function finite at each point $x \in R^n$;

(2) *there exists $\mu^* \in M$ such that for any motion $x(t;t_0,x_0,\mu)$ of the system (2.2.2) the function*

$$v_0(t, x(t;t_0,x_0,\mu)) + u(x(t;t_0,x_0,\mu),\mu)$$

is not increscent at all $t \geq t_0$ and $\mu < \mu^*$.

Then the motion $x(t,\mu)$ of the system (2.2.2) is μ-bounded uniformly with respect to t_0 with respect to the subvector $y = (x_1^T,\ldots,x_k^T)^T$.

Proof. Sufficiency For any $x_0 \in R^n$ choose a value $N(x_0) > 0$ so that at $\|y\| > N(x_0)$ the inequality $a(\|y\|) > w(x_0)$ would hold.
According to conditions (1) and (2) of Theorem 2.5.2, obtain

$$a(\|y(t,\mu)\|) \leq v_0(t, x(t;t_0,x_0,\mu)) + u(x(t;t_0,x_0,\mu),\mu)$$
$$\leq v_0(t_0,x_0) + u(x_0,\mu) \leq w(x_0) \leq a(N)$$

at all $t \geq t_0$ and $\mu < \mu^*$. Hence $\|y(t,\mu)\| < N(x_0)$ at all $t \geq t_0$ and $\mu < \mu^*$.

Necessity If the motion $x(t,\mu)$ of the system (2.2.2) is μ-bounded uniformly with respect to t_0 with respect to the subvector $y = (x_1^T,\ldots,x_k^T)^T$, then the function (2.5.4) is defined in the domain $R_+ \times R^n$. In addition,

$$v_0(t,x) + u(x,\mu) \leq N(x) \equiv w(x) \quad at \quad \mu < \mu^*.$$

This function is not increscent along solutions of the system (2.2.2) at all $t \geq t_0$ and $\mu < \mu^*$.

Thus, Theorem 2.5.2 is proved.

Theorem 2.5.3 *Assume that the motion equations (2.2.2) are such that:*

(1) *the strengthened function $v_0(t, x) + u(x, \mu)$ in the range of values $(t, x) \in R_+ \times R^n$ satisfies the condition (2.5.3) and for any compact set $E \subset R^n$ there exists a function $\Delta_E(t)$ such that*

$$v_0(t, x) + u(x, \mu) \leq \Delta_E(t) \quad at \quad x \in E, t \geq 0, \mu < \mu^* \in M;$$

(2) *there exists $\mu^* \in M$ such that for any motion $x(t; t_0, x_0, \mu)$ the function*

$$v_0(t, x(t; t_0, x_0, \mu)) + u(x(t; t_0, x_0, \mu), \mu)$$

is not increscent at all $t \geq t_0$ and $\mu < \mu^*$.

Then the motion $x(t, \mu)$ of the system (2.2.2) is μ-bounded with respect to x_0 with respect to the subvector $y = (x_1^{\mathrm{T}}, \ldots, x_k^{\mathrm{T}})^{\mathrm{T}}$.

Proof. Sufficiency According to the condition (2.5.3) for any $t_0 \geq 0$ and a compact set E there exists $N(t_0, E) > 0$ and $\mu \in M$ such that:

(a) at $\|y\| > N(t_0, E)$ the following inequality holds:

$$a(\|y\|) > \varphi_E(t_0);$$

(b) at $x_0 \in E$, $t \geq t_0$ and $\mu < \mu^*$ the following estimates hold:

$$a(\|y(t; t_0, x_0, \mu)\|) \leq v_0(t, x(t; t_0, x_0, \mu)) + u(x(t; t_0, x_0, \mu), \mu)$$
$$\leq v_0(t_0, x_0) + u(x_0, \mu) \leq \Delta_E(t_0) \leq a(N).$$

Hence find $\|y(t; t_0, x_0, \mu)\| \leq N$ at $t \geq t_0$ and $\mu < \mu^*$.

Necessity The function $v_0(t, x) + u(x, \mu)$ determined by the formula (2.5.4) satisfies the estimate

$$v_0(t, x) + u(x, \mu) \leq N(t, E) \equiv \varphi_E(t), \quad \mu < \mu^*,$$

and is not increscent along the solutions $x(t; t_0, x_0, \mu)$ of the system (2.2.2).

Theorem 2.5.3 is proved.

Theorem 2.5.4 *Assume that the motion equations (2.2.2) are such that:*

(1) *the strengthened function $v_0(t, x) + u(x, \mu)$ in the range of values $(t, x) \in R_+ \times R^n$ satisfies the condition (2.5.3) and, in addition, there exists a function b: $b(r) \to +\infty$ at $r \to +\infty$ such that*

$$v_0(t, x) + u(x, \mu) \leq b(\|x\|);$$

(2) *there exists $\mu^* \in M$ such that for any motion $x(t; t_0, x_0, \mu)$ of the system (2.2.2) the function*

$$v_0(t, x(t; t_0, x_0, \mu)) + u(x(t; t_0, x_0, \mu), \mu)$$

is not increscent at all $t \geq t_0$ and $\mu < \mu^*$.

Then the motion $x(t, \mu)$ of the system (2.2.2) is μ-bounded uniformly with respect to (t_0, x_0) with respect to the subvector $y = (x_1^{\mathrm{T}}, \ldots, x_k^{\mathrm{T}})^{\mathrm{T}}$.

Proof. Sufficiency For any compact set $E \subset R^n$ calculate the value

$$b_E = \sup[v_0(t, x) + u(x, \mu)\colon\ t \geq 0, x \in E] \leq \sup[b(\|x\|)\colon\ x \in E] < +\infty.$$

From conditions (1) and (2) of Theorem 2.5.4 it follows that there exists $N(E) > 0$ and $\mu^* \in M$ such that:

(a) at $\|y\| > N(E)$ the inequality holds true:

$$a(\|y\|) > b_E;$$

(b) at $\mu < \mu^*$, $t_0 \geq 0$ and $x_0 \in E$ the following estimate holds:

$$a(\|y(t; t_0, x_0, \mu)\|) \leq v_0(t, x(t; t_0, x_0, \mu)) + u(x(t; t_0, x_0, \mu), \mu)$$
$$\leq v_0(t_0, x_0) + u(x_0, \mu) \leq b_E \leq a(N).$$

Hence, $\|y(t; t_0, x_0, \mu)\| \leq N$ for all $t \geq t_0$ and $\mu < \mu^*$.

Necessity If the motion $x(t, \mu)$ of the system (2.2.2) is μ-bounded uniformly with respect to $\{t_0, x_0\}$ with respect to the subvector

$$y = (x_1^{\mathrm{T}}, \ldots, x_k^{\mathrm{T}})^{\mathrm{T}},$$

then there exists a function $v_0(t, x) + u(x, \mu)$ determined by the formula (2.5.4). For compact sets E use the balls $\|x\| = r$, $r \in [0, \infty)$, and for the values $(t, x) \in R_+ \times R^n$ obtain

$$v_0(t, x) + u(x, \mu) \leq N(E) = N(r)$$
$$\text{at}\quad \mu < \mu^*,\quad E \subset R^n,\quad r \in [0, \infty).$$

The function $N(r) \to +\infty$ at $r \to +\infty$; therefore, one can assume $b(\|x\|) = N(\|x\|)$. The function $v_0(t, x) + u(x, \mu)$ determined by the formula (2.5.4) is not increscent along solutions of the system (2.2.2).

Theorem 2.5.4 is proved.

Remark 2.5.1 If the strengthened function

$$v_0(t, x) + u(x, \mu),\quad \mu < \mu^*,$$

has continuous first-order partial derivatives, then the condition for the non-increase of the function

$$v_0(t, x(t; t_0, x_0, \mu)) + u(x(t; t_0, x_0, \mu), \mu)$$

along solutions of the system (2.2.2) can be substituted by the condition

$$\frac{d}{dt}(v_0(t, x) + u(x, \mu))|_{(2.2.2)} \leq 0$$

in the range of values $(t, x) \in R_+ \times R^n$ at $\mu < \mu^*$.

Remark 2.5.2 If in the system (2.2.2) we assume that $\mu = 0$, $x_s = x \in R^n$ and $s = 1$, then Theorems 2.5.1–2.5.4 imply the statements of Theorem 39.1 from the monograph by Rumiantsev and Oziraner [1].

2.6 Algebraic Conditions of μ-Boundedness

Now for the study of μ-boundedness of motion of the system (2.2.2), construct the algebraic necessary conditions, using the functions $v_s(t, x_s)$, $s = 1, 2, \ldots, m$ only, which were constructed for the independent subsystems (2.2.4). Those conditions will be based on the following assumption on the independent subsystems.

Assumption 2.6.1 There exist:

(1) continuously differentiable functions $v_s(t, x_s)$, $v_s \colon R_+ \times R^{n_s} \to R_+$, $s = 1, 2, \ldots, m$;

(2) comparison functions $\psi_{s1}, \psi_{s2}, \psi_{s3}$ from the KR-class, $s = 1, 2, \ldots, m$;

(3) constants $\sigma_s \in R$, $s = 1, 2, \ldots, m$, such that

    (a) $\psi_{s1}(\|x_s\|) \leq v_s(t, x_s) \leq \psi_{s2}(\|x_s\|)$

    (b) $\left.\dfrac{dv_s(t, x_s)}{dt}\right|_{(2.2.4)} \leq \sigma_s \psi_{s3}(\|x_s\|)$

    at all $t \in R_+$ and all $\|x_s\| \geq r_s$ (r_s may be sufficiently large);

(4) the functions $v_s(t, x_s)$ and $\left.\dfrac{dv_s}{dt}\right|_{(2.2.4)}$ at all $s = 1, 2, \ldots, m$ are bounded on the sets $R_+ \times S(r_s)$, $s = 1, 2, \ldots, m$.

Theorem 2.6.1 *Let the motion equations (2.2.1) with the decomposition (2.2.2) be such that:*

(1) *all the conditions of Assumption 2.6.1 are satisfied;*

(2) *at the specified functions* $v_s \colon R_+ \times R^{n_s} \to R_+$ *and* ψ_{s3} *from the KR-class there exist constants* $a_{sj} \in R$ *such that*

$$\left(\frac{\partial v_s(t, x_s)}{\partial x_s}\right)^{\mathrm{T}} g_s(t, x_1, \ldots, x_m) \le [\psi_{s3}(\|x_s\|)]^{1/2} \sum_{j=1}^{m} a_{sj} [\psi_{s3}(\|x_j\|)]^{1/2}$$

at all $x_s \in R^{n_s}$, $x_j \in R^{n_s}$, $s = 1, 2, \ldots, m$ *and* $t \in R_+$;

(3) *at the specified constants* $\sigma_s \in R$ *there exists a value of the parameter* $\mu^* \in M$ *and an m-vector* $a^{\mathrm{T}} = (a_1, \ldots, a_m)$ *such that the matrix* $S(\mu) = [s_{ij}(\mu)]$ *with the elements*

$$s_{ij}(\mu) = \begin{cases} a_s(\sigma_s + \mu a_{ss}), & s = j, \\ \dfrac{1}{2}\mu(a_j a_{sj} + a_j a_{js}), & s \ne j, \end{cases}$$

is negative semidefinite (definite) at $\mu \in (0, \mu^*]$ *and at* $\mu \to 0$.

Then the motion $x(t, \mu)$ *of the system (2.2.2) is uniformly μ-bounded (uniformly ultimately μ-bounded).*

Proof Using the functions $v_s(t, x_s)$ and the vector $a = (a_1, \ldots, a_m)^{\mathrm{T}}$ construct the scalar function

$$v(t, x, a) = \sum_{s=1}^{m} a_s v_s(t, x_s). \tag{2.6.1}$$

According to condition (3a) from Assumption 2.6.1, for the function (2.6.1) the following estimates hold:

$$\psi_1(\|x\|) \le v(t, x, a) \le \psi_2(\|x\|), \tag{2.6.2}$$

where $\psi_1, \psi_2 \in KR$-class. In addition, condition (3b) and condition (2) of Theorem 2.6.1 at all $t \in R$ imply the estimate

$$\sum_{s=1}^{m} a_s \left[\frac{dv_s(t, x_s)}{dt}\right]_{(2.2.2)} \le \lambda_M(\mu) \psi_3(\|x\|), \tag{2.6.3}$$

where $\lambda_M(\mu) < 0$, as soon as $x \in R^n \setminus (S_1(r_1) \times \ldots \times S_m(r_m))$. Here the comparison function $\psi_3 \in KR$-class.

Now we will consider the situation where $\|x_i\| \ge r_i$ for $i = 1, 2, \ldots, l$, $l < m$ and $\|x_i\| < r_i$ at $i = l+1, \ldots, m$. For the function (2.6.1) consider the

estimates

$$\sum_{s=1}^{l} a_s \psi_{s1}(\|x_s\|) + \sum_{s=l+1}^{m} a_s v_s(t, x_s) \le v(t, x, a) \le$$

$$\le \sum_{s=1}^{l} a_s \psi_{s1}(\|x_s\|) + \sum_{s=l+1}^{m} a_s v_s(t, x_s). \tag{2.6.4}$$

The fact that $v_s(t, x_s)$ are continuous on $R_+ \times R^{n_s}$ and bounded on $R_+ \times S_s(r_s)$, $s = 1, 2, \ldots, l$, implies the existence of comparison functions φ_1, φ_2 from the KR-class, such that

$$\varphi_1(\|x\|) \le v(t, x, a) \le \varphi_2(\|x\|) \tag{2.6.5}$$

at all $t \in R_+$ and all $x \in R^n$, while $\|x_s\| < r_s$, $s = l+1, \ldots, m$, and the values $\|x_s\|$ are sufficiently large for $s = 1, 2, \ldots, l$.

For the function $\dfrac{dv(t, x, a)}{dt}$ along solutions of the system (2.2.2) obtain

$$\frac{dv(t, x, a)}{dt}\bigg|_{(2.2.2)} = \sum_{s=1}^{l} a_s \left\{ \left[\frac{dv_s(t, x_s)}{dt} \right]_{(2.2.4)} \right.$$

$$+ \left. \left(\frac{\partial v_s(t, x_s)}{\partial x_s} \right)^{\mathrm{T}} \mu g_s(t, x_1, \ldots, x_m) \right\}$$

$$+ \sum_{s=l+1}^{m} a_s \left\{ \frac{dv_s(t, x_s)}{dt}\bigg|_{(2.2.4)} + \left(\frac{\partial v_s(t, x_s)}{\partial x_s} \right)^{\mathrm{T}} \mu g_s(t, x_1, \ldots, x_m) \right\}$$

$$\le \sum_{s=1}^{l} a_s \sigma_s \psi_{s3}(\|x_s\|) + \sum_{s=1}^{l} a_s [\psi_{s3}(\|x_s\|)]^{\frac{1}{2}} \mu \sum_{l=1}^{l} a_{sj} [\psi_{j3}(\|x_j\|)]^{\frac{1}{2}}$$

$$+ \sum_{s=1}^{l} a_s [\psi_{s3}(\|x_s\|)]^{\frac{1}{2}} \mu \sum_{j=l+1}^{m} a_{sj} [\psi_{j3}(\|x_j\|)]^{\frac{1}{2}} \tag{2.6.6}$$

$$+ \sum_{s=l+1}^{m} a_s \frac{dv_s(t, x_s)}{dt}\bigg|_{(2.2.4)}$$

$$+ \sum_{s=l+1}^{m} a_s [\psi_{s3}(\|x_s\|)]^{\frac{1}{2}} \mu \sum_{j=1}^{l} a_{sj} [\psi_{j3}(\|x_j\|)]^{\frac{1}{2}}$$

$$+ \sum_{s=l+1}^{m} a_s [\psi_{s3}(\|x_s\|)]^{\frac{1}{2}} \mu \sum_{j=l+1}^{m} a_{sj} [\psi_{j3}(\|x_j\|)]^{\frac{1}{2}}.$$

For all $\|x_s\| < r_s$, $s = l+1, \ldots, m$, there exist constants K, K_1, K_2, K_3, K_{4s}, $s = l+1, \ldots, m$, such that

$$\sum_{j=l+1}^{m} |a_{sj}| [\psi_{j3}(\|x_j\|)]^{\frac{1}{2}} \le K_1, \tag{2.6.7}$$

$$\sum_{i=l+1}^{m} a_i [\psi_{i3}(\|x_i\|)]^{\frac{1}{2}} \le K_2, \qquad (2.6.8)$$

$$\sum_{s=l+1}^{m} a_s [\psi_{s3}(\|x_s\|)]^{\frac{1}{2}} \sum_{j=1}^{l} a_{sj} [\psi_{j3}(\|x_j\|)]^{\frac{1}{2}} \le K_3, \qquad (2.6.9)$$

$$\left| \frac{dv_s(t, x_s)}{dt} \right|_{(2.2.2)} \le K_{4s}, \quad s = l+1, \dots, m. \qquad (2.6.10)$$

Let $w = (\psi_{13}(\|x_1\|), \dots, \psi_{l3}(\|x_l\|))^{\mathrm{T}}$ and let $P = [p_{ij}]$ be an $l \times l$-matrix with the elements

$$p_{ij}(\mu) = \begin{cases} a_i(\sigma_i + \mu a_{ii}), & i = j, \\ \mu a_i a_{ij}, & i \ne j. \end{cases}$$

Denote $\tilde{S} = \frac{1}{2}(P + P^{\mathrm{T}})$. Then the estimate of the expression (2.6.6), taking into account the inequalities (2.6.7)–(2.6.10), has the form

$$\frac{dv(t, x, a)}{dt} \bigg|_{(2.2.2)} \le w^{\mathrm{T}} \tilde{S} w + \mu K_1 \sum_{s=1}^{l} a_s [\psi_{s3}(\|x_s\|)]^{\frac{1}{2}}$$

$$+ \sum_{s=l+1}^{m} a_s K_{4s} + \mu K_2 \sum_{j=1}^{l} |a_{sj}| [\psi_{j3}(\|x_j\|)]^{\frac{1}{2}} + \mu K_3.$$

According to condition (3) of Theorem 2.6.1, there exists $\mu^* \in M$ such that the matrix $\tilde{S}(\mu)$ is negative semidefinite (definite) at $\mu < \mu^*$. Therefore, $\lambda_M(\tilde{S}) < 0$ and then

$$\frac{dv(t, x, a)}{dt} \bigg|_{(2.2.2)} \le \lambda_M(\tilde{S}) \sum_{i=l}^{l} \psi_{i3}(\|x_i\|) + \mu K_1 \sum_{s=l}^{r} a_s [\psi_{s3}(\|x_s\|)]^{\frac{1}{2}}$$

$$+ \sum_{s=l+1}^{m} a_s K_{4s} + \mu K_2 \sum_{j=l}^{l} |a_{sj}| [\psi_{j3}(\|x_j\|)]^{\frac{1}{2}} + \mu K_3. \qquad (2.6.11)$$

Since $\lambda_M(\tilde{S}(\mu)) < 0$, at any value of $\|x_s\|$, $s = 1, 2, \dots, l$, one can find $\mu^{**} \in M$ such that the sign of $\dfrac{dv(t, x, a)}{dt} \bigg|_{(2.2.2)}$ will be determined by the expression

$$\lambda_M(\tilde{S}) \sum_{i=l}^{l} \psi_{i3}(\|x_i\|)$$

at $\mu < \mu^{**}$. Thus, at $\mu < \mu^0 = \min(\mu^*, \mu^{**})$ the function $v(t, x, a)$ is positive definite and decrescent and its full derivative (2.6.11) is negative definite. According to Theorems 10.4 and 10.5 from the monograph of Yoshizawa [2], the state $x = 0$ of the system (2.2.2) is uniformly μ-bounded (uniformly ultimately μ-bounded).

Theorem 2.6.1 is proved.

Remark 2.6.1 Along with the use of the function (2.6.1) for the analysis of μ-boundedness of motion of the system (2.2.2) it is possible to use the vector function

$$v(t, x) = (v_1(t, x_1), \ldots, v_m(t, x_m))^{\mathrm{T}}$$

and the theory of M-matrices.

Recall some definitions, following the monographs of Michel and Miller [1] and Šiljak [1].

Definition 2.6.1 A real $(m \times m)$-matrix $D = [d_{sj}]$ is called an M-matrix if $d_{sj} \leq 0$, $s \neq j$ (i.e., all off-diagonal elements of the matrix D are not positive) and all principal minors of the matrix D are positive.

Definition 2.6.2 The real $(m \times m)$-matrix $A = [a_{sj}]$ is called a matrix with the dominant main diagonal, if there exist positive numbers d_j, $j = 1, 2, \ldots, s$, such that

$$d_s|a_{ss}| > \sum_{j=1, j\neq s}^{m} d_j|a_{sj}| \quad \text{for all} \quad s = 1, 2, \ldots, m$$

or

$$d_j|a_{jj}| > \sum_{s=1, s\neq j}^{m} d_s|a_{sj}| \quad \text{for all} \quad j = 1, 2, \ldots, m.$$

Consider the following statement.

Theorem 2.6.2 *Let the motion equations (2.2.1) with the decomposition (2.2.2) be such that:*

(1) *all the conditions of Assumption 2.6.1 are satisfied;*

(2) *condition (2) of Theorem 2.6.1 is satisfied with the constants $a_{sj} \geq 0$ at $s \neq j$;*

(3) *at the specified constants $\sigma_s \in R$ there exists a value of the parameter $\mu^* \in M$ such that all the main diagonal minors of the matrix $D(\mu) = [d_{sj}(\mu)]$ are positive at $\mu < \mu^*$, where*

$$d_{sj}(\mu) = \begin{cases} -(\sigma_s + \mu a_{ss}), & s = j, \\ -\mu a_{sj}, & s \neq j. \end{cases}$$

Then the motion $x(t, \mu)$ of the system (2.2.2) is uniformly μ-bounded (uniformly ultimately μ-bounded).

The proof of Theorem 2.6.2 is similar to that of Theorem 2.6.1 and therefore is not given here.

2.7   Applications

2.7.1   Lienard oscillator

The Lienard equation is one of the important differential equations widely used in mechanics and electrical engineering (see Česari [1], Burton [2], Reissig et al. [1], and others). The study of solutions of this equation is still the focus of attention for a lot of specialists, which is witnessed by numerous publications in academic periodicals.

Study the μ-boundedness of solutions of the equation

$$\ddot{x} + f(t, x, \dot{x}, \mu)\dot{x} + g(x) = 0, \qquad (2.7.1)$$

where $f\colon R_+ \times R \times R \times M \to R_+$, $g\colon R \to R$, the functions f and g are continuous, $f(t, x, y, \mu) > 0$, if $y \neq 0$ and $\mu \in M^0 \subset M$, and $xg(x) > 0$, if $x \neq 0$. The equation (2.7.1) is equivalent to the system

$$\begin{aligned} \dot{x} &= y, \\ \dot{y} &= -f(t, x, y, \mu)y - g(x). \end{aligned} \qquad (2.7.2)$$

Choose the function

$$v(x, y) = W(x) + \frac{1}{2}y^2,$$

where $W(x) = \int\limits_0^x g(s)\, ds$. Then obtain

$$Dv(x, y)|_{(2.7.2)} = -f(t, x, y, \mu)y^2 \leq 0 \qquad (2.7.3)$$

at all $(t, x, y, \mu) \in R_+ \times R \times R \times M^0$.

Assume that $W(-\infty)$ and $W(+\infty)$ are finite quantities and determine the sets

$$\begin{aligned} S_1 &= \{(x, y)\colon\ x > 0\}, \\ S_2 &= \{(x, y)\colon\ x < 0\}. \end{aligned}$$

Take the strengthening functions $u_i(x, y, \mu)$, $i = 1, 2$, in the form

$$\begin{aligned} u_1(x, y, \mu) &= \mu x, \\ u_2(x, y, \mu) &= -\mu x, \quad \mu \in M^0, \end{aligned} \qquad (2.7.4)$$

and assume that $J = 1$. Let $X = (x, y)^{\mathrm{T}}$ and denote the right-hand part of the system (2.7.2) by $F(t, X, \mu) = (y, -f(t, x, y, \mu)y - g(x))^{\mathrm{T}}$. Take L_1 and L_2 in the form $L_1 = \Phi(+\infty)$ and $L_2 = \Phi(-\infty)$. Note that

$$|\operatorname{grad} u_i(x, y, \mu)F(t, X, \mu)| = |y|,$$

$$v(t, X) - L_1 = W(x) - W(+\infty) + \frac{1}{2}y^2,$$

$$v(t, X) - L_2 = W(x) - W(-\infty) + \frac{1}{2}y^2.$$

Then if $u_i(x, y, \mu) \geq J$, then $v(t, x) - L_i < \frac{1}{2}y^2$.

If the function $\Phi\colon (0, L - L_i) \to R_+$ from Definition 2.4.3 is taken in the form $\Phi(s) = (2s)^{1/2}$, then at $u_i(x, y, \mu) \geq J$ and $v(t, x, y) > L_i$ obtain

$$\Phi(v(t, x, y) - L_i) < |y| \quad \text{at} \quad i = 1, 2.$$

Now, let $L > L_1$ and a function $h(x)$ such that $\int\limits_J^\infty h(s)\, ds = +\infty$ be specified. If $L \geq v(t, x, y) > L_1$ and $u_1(x, y, \mu) \geq J$, then, choosing $\mu \in M^1 \subset M$, one can obtain

$$f(t, x, y, \mu) \geq h(x) \geq 0 \quad \text{and} \quad \int\limits_J^\infty h(s)\, ds = +\infty. \tag{2.7.5}$$

Similarly, for a specified $L > L_2$, if $L \geq v(t, x, y) > L_2$ and $u_2(t, y, \mu) \geq J$, then, choosing $\mu \in M^2 \subset M$, one can obtain

$$f(t, x, y, \mu) \geq h(x) \geq 0 \quad \text{and} \quad \int\limits_{-J}^{-\infty} h(s)\, ds = -\infty. \tag{2.7.6}$$

Under the conditions (2.7.5) and (2.7.6) for $J \leq u_i(x, y, \mu)$, $i = 1, 2$, and $L \geq v(t, x, y) > L_i$ obtain

$$Dv(t, x, y)|_{(2.7.2)} \leq \Phi(v(t, x, y) - L_i))h(u_i(x, y, \mu))|Du_i(x, y, \mu)|_{(2.7.2)} \tag{2.7.7}$$

at $\mu \in M^0 = M^1 \cap M^2$.

Thus, all the conditions of Theorem 2.4.3 are satisfied and solutions of the system (2.7.2) are μ-bounded.

Remark 2.7.1 If $f(t, x, y, \mu) = h(x)$ at all $(t, x, y, \mu) \in R_+ \times R \times R \times M$, then the conditions (2.7.5) and (2.7.6) are necessary and sufficient for the boundedness of solutions of the system (2.7.2) according to the results of the article of Burton [2].

2.7.2 Connected systems of Lurie–Postnikov equations

Consider an indirect control system

$$
\begin{aligned}
\frac{dx_1}{dt} &= A_1 x_1 + b_1 f(\sigma), \\
\frac{dx_2}{dt} &= A_2 x_2 + b_2 f(\sigma), \\
\frac{d\sigma}{dt} &= \mu c_1^{\mathrm{T}} x_1 + \mu c_2^{\mathrm{T}} x_2 - r f(\sigma),
\end{aligned}
\tag{2.7.8}
$$

where $x_1 \in R^{n_1}$, $x_2 \in R^{n_2}$, A_1 is an $(n_1 \times n_1)$-matrix, A_2 is an $(n_2 \times n_2)$-matrix, $b_1 \in R^{n_1}$, $b_2 \in R^{n_2}$, $c_1 \in R^{n_1}$, $c_2 \in R^{n_2}$, $n_1 + n_2 = n$, $f\colon R \to R$, $\sigma f(\sigma) > 0$, if $\sigma \neq 0$, f is a continuous function, $\mu \in M$.

Define the conditions for the μ-boundedness of motion of the system (2.7.8) on the basis of Theorem 2.4.3. For this purpose, consider the two functions

$$V_1(x_1,\sigma) = x_1^T B_1 x_1 + W(\sigma), \qquad (2.7.9)$$
$$V_2(x_2,\sigma) = x_2^T B_2 x_2 + W(\sigma), \qquad (2.7.10)$$

where $W(\sigma) = \int_0^\sigma f(\sigma)\,d\sigma$, B_1 and B_2 are positive definite matrices of the dimensions $n_1 \times n_1$ and $n_2 \times n_2$, respectively. For the function

$$V_0(x_1,x_2,\sigma) = V_1(x_1,\sigma) + V_2(x_2,\sigma) \qquad (2.7.11)$$

obtain

$$\left.\frac{dV_0(x_1,x_2,\sigma)}{dt}\right|_{(2.7.8)} = -x_1^T D_1 x_1 - x_2^T D_2 x_2$$
$$+ f(\sigma)\big[2b_1^T B_1 + \mu c_1^T\big]x_1 \qquad (2.7.12)$$
$$+ f(\sigma)\big[2b_2^T B_2 + \mu c_2^T\big]x_2 - 2rf(\sigma).$$

The function $dV_0(x_1,x_2,\sigma)/dt|_{(2.7.8)}$ will be negative definite if

$$D_1 = -(A_1^T B_1 + B_1 A_1),$$
$$D_2 = -(A_2^T B_2 + B_2 A_2) \qquad (2.7.13)$$

and

$$r > \min\{r_1,r_2\}, \qquad \mu^0 = \min\{\mu_1^0,\mu_2^0\}, \qquad (2.7.14)$$

where

$$r_1 > (B_1 b_1 + \mu c_1/2)^T D_1^{-1}(B_1 b_1 + \mu c_1/2), \qquad \mu < \mu_1^0 \in M,$$
$$r_2 > (B_2 b_2 + \mu c_2/2)^T D_2^{-1}(B_2 b_2 + \mu c_2/2), \qquad \mu < \mu_2^0 \in M.$$

Let $M = 0$ (see condition (b) in Theorem 2.4.2) and let $W(\pm\infty) \neq \infty$. Here the function $V_0(x_1,x_2,\sigma)$ is not radially unbounded and therefore its application for the analysis of the μ-boundedness of the system (2.7.8) is impossible.

Now assume that $W(\infty) = W(-\infty)$ and choose $L_1 = L_2 = W(\infty)$. Define the surfaces $S_1,\ldots,S_4$ as follows:

$$S_1 = \{(x_1,\sigma)\colon \sigma > 0\}, \quad S_2 = \{(x_1,\sigma)\colon \sigma < 0\},$$
$$S_3 = \{(x_2,\sigma)\colon \sigma > 0\}, \quad S_4 = \{(x_2,\sigma)\colon \sigma < 0\}. \qquad (2.7.15)$$

Choose the strenghthening function on the basis of the conditions

$$u_1(x_1,\sigma) = \sigma, \quad u_2(x_1,\sigma) = -\sigma,$$
$$u_3(x_2,\sigma) = \sigma, \quad u_4(x_2,\sigma) = -\sigma. \qquad (2.7.16)$$

Under (2.7.14), from the negative definiteness of the function $\frac{dV_0(x_1,x_2,\sigma)}{dt}\big|_{(2.7.8)}$ it follows that there exists a constant $m > 0$ such that

$$\frac{dV_0(x_1,x_2,\sigma)}{dt}\bigg|_{(2.7.8)} \leq -m(x_1^{\mathrm{T}} x_1 + x_2^{\mathrm{T}} x_2 + f^2(\sigma)). \qquad (2.7.17)$$

Now

$$V_0(x_1,x_2,\sigma) - W(\infty) = x_1^{\mathrm{T}} B_1 x_1 + x_2^{\mathrm{T}} B_2 x_2 + W(\sigma) - W(\infty)$$
$$\leq x_1^{\mathrm{T}} B_1 x_1 + x_2^{\mathrm{T}} B_2 x_2 \leq Q(x_1^{\mathrm{T}} x_1 + x_2^{\mathrm{T}} x_2)$$

for some $Q > 0$.

In addition,

$$|\sigma'| = |\mu c_1^{\mathrm{T}} x_1 + \mu c_2^{\mathrm{T}} x_2 - r f(\sigma)| \leq P[\mu(x_1^{\mathrm{T}} x_1 + x_2^{\mathrm{T}} x_2) + f^2(\sigma)]^{1/2}$$

for $\mu < \mu^* \in M$ and some $P > 0$.

Assuming $h(s) = m/\mu P Q^{1/2}$, obtain

$$\frac{dV_0(x_1,x_2,\sigma)}{dt}\bigg|_{(2.7.11)} \leq -(m/\mu P Q^{1/2})[V_0(x_1,x_2,\sigma) - W(\infty)]^{1/2}|\dot{u}_i| \quad (2.7.18)$$

at $\mu < \mu^*$ and $i = 1, 2, 3, 4$.

According to Theorem 2.4.3, every motion of the weakly connected system (2.7.8) is μ-bounded.

2.7.3 A nonlinear system with weak linear connections

Consider the linear system with weak linear connections

$$\begin{aligned}\frac{dx_1}{dt} &= f_1(t, x_1) + \mu C_{12} x_2, \\ \frac{dx_2}{dt} &= f_2(t, x_2) + \mu C_{21} x_1,\end{aligned} \qquad (2.7.19)$$

where $x_i \in R^{n_i}$, $i = 1, 2$, $f_i : R_+ \times R^{n_i}$, C_{ij} are matrices, μ is a small parameter. At $\mu = 0$ the system (2.7.19) falls into two independent nonlinear subsystems

$$\frac{dx_1}{dt} = f_1(t, x_1), \quad x_1(t_0) = x_{10}, \qquad (2.7.20)$$

$$\frac{dx_2}{dt} = f_2(t, x_2), \quad x_2(t_0) = x_{20}. \qquad (2.7.21)$$

Introduce the following assumption for the systems (2.7.20) and (2.7.21).

Assumption 2.7.1 For the independent subsystems (2.7.20) and (2.7.21) there exist:

(1) functions $v_1(t, x_1)$ and $v_2(t, x_2)$, continuous and continuously differentiable on $R_+ \times R^{n_i}$, $i = 1, 2$;

(2) constants $c_{11}, \ldots, c_{15}$ and $c_{21}, \ldots, c_{25}$ such that

    (a) $-c_{11}\|x_1\|^2 \leq v_1(t, x_1) \leq -c_{12}\|x_1\|^2$,

    (b) $-c_{14}\|x_1\|^2 \leq \left.\dfrac{dv_1}{dt}\right|_{(2.7.20)} \leq -c_{13}\|x_1\|^2$,

    (c) $\left\|\dfrac{\partial v_1}{\partial x_1}(t, x_1)\right\| \leq -c_{15}\|x_1\|^2$

at all $t \in R_+$ and $x_1 \in R^{n_1}$;

    (a') $c_{21}\|x_2\|^2 \leq v_2(t, x_2) \leq c_{22}\|x_2\|^2$,

    (b') $\left.\dfrac{dv_2}{dt}\right|_{(2.7.21)} \leq -c_{23}\|x_2\|^2$,

    (c') $\left\|\dfrac{\partial v_2}{\partial x_2}(t, x_2)\right\| \leq c_{25}\|x_2\|^2$

at all $t \in R_+$ and $x_2 \in R^{n_2}$.

Taking into account that

$$\left(\frac{\partial v_1}{\partial x_1}(t, x_1)\right)^{\mathrm{T}} g_1(t, x) = \left(\frac{\partial v_1}{\partial x_1}(t, x_1)\right)^{\mathrm{T}} C_{12} x_2, \qquad (2.7.22)$$

$$\left(\frac{\partial v_2}{\partial x_2}(t, x_2)\right)^{\mathrm{T}} g_2(t, x) = \left(\frac{\partial v_2}{\partial x_2}(t, x_1)\right)^{\mathrm{T}} C_{21} x_1, \qquad (2.7.23)$$

find the estimates

$$\left\|\left(\frac{\partial v_1}{\partial x_1}\right)^{\mathrm{T}} C_{12} x_2\right\| \leq \left\|\frac{\partial v_1}{\partial x_1}\right\| \|C_{12}\| \|x_2\| \leq c_{15}\|C_{12}\| \|x_1\| \|x_2\|,$$

$$\left\|\left(\frac{\partial v_2}{\partial x_2}\right)^{\mathrm{T}} C_{21} x_1\right\| \leq \left\|\frac{\partial v_2}{\partial x_2}\right\| \|C_{21}\| \|x_1\| \leq c_{25}\|C_{21}\| \|x_1\| \|x_2\|$$

in the domain $x_1 \in R^{n_1}$, $x_2 \in R^{n_2}$.

According to condition (3) of Theorem 2.6.1, elements of the matrix $S(\mu)$ have the form

$$s_{11} = -a_1 c_{13}, \quad s_{22} = -a_2 c_{23},$$
$$s_{12} = s_{21} = 1/2[\mu(a_1 c_{15}\|C_{12}\| + s_2 c_{25}\|C_{21}\|)].$$

Choose a vector $a = (a_1, a_2)^{\mathrm{T}} > 0$ with the components

$$a_1 = \frac{1}{c_{15}\|C_{12}\|}, \quad a_2 = \frac{1}{c_{25}\|C_{21}\|}$$

and assume that the function

$$V(t, x) = a_1 V_1(t, x_1) + a_2 V_2(t, x_2) \qquad (2.7.24)$$

satisfies condition (2) of Theorem 2.6.1.

The matrix $S(\mu)$ from condition (3) of the theorem has the form

$$S(\mu) = \begin{bmatrix} -\dfrac{c_{13}}{c_{15}\|C_{12}\|} & \mu \\ \mu & -\dfrac{c_{23}}{c_{25}\|C_{21}\|} \end{bmatrix}.$$

If the conditions

$$-\frac{c_{13}}{c_{15}\|C_{12}\|} < 0, \quad \mu^2 \|C_{12}\|\|C_{21}\| < \frac{c_{13}c_{23}}{c_{15}c_{25}}$$

are satisfied at all $\mu \in (0, \mu^*]$, where

$$\mu^* = \left[\frac{c_{13}c_{23}}{c_{15}c_{25}\|C_{12}\| \, \|C_{21}\|} \right]^{1/2},$$

then the matrix $S(\mu)$ at $\mu \in (0, \mu^*]$ is negative definite. Now Corollary 2.3.12 of Theorem 2.3.3 according to which the motion of the weakly connected system (2.7.19) is uniformly ultimately μ-bounded can be applied to the function (2.7.24) and the estimate

$$\frac{dV_0}{dt} \le u^{\mathrm{T}} S(\mu) u, \qquad (2.7.25)$$

where $u = (\|x_1\|, \|x_2\|)^{\mathrm{T}}$.

2.8 Comments and References

2.2. The statement of the problem of the boundedness of motion of systems with a small parameter is formulated, taking into account the known results (see Česari [1], Yoshizawa [1, 2], Pliss [1], Lakshmikantham, Leela, and Martynyuk [1]).

2.3. Theorems 2.3.1–2.3.3 are new. To obtain them, strengthened Lyapunov functions (see Burton [1]) and two measures (see Lakshmikantham and Salvadori [1], Movchan [1]) are applied. Under some special assumptions, the obtained results imply the known results obtained for systems that do not contain a small parameter (cf. Reissig, Sansone, and Conti [1], Yoshizawa [2], Lakshmikantham and Liu [1]).

2.4. The comparison technique is applied, the basic ideas were stated in the monograph by Lakshmikantham, Leela, and Martynyuk [1] (see Lemmas 2.4.1–2.4.3). Theorem 2.4.1 is taken from the work of Mitropolsky and V.A. Martynyuk [1]. Theorems 2.4.2 and 2.4.3 are new.

2.5. For the investigation of the boundedness of weakly connected equations with respect to a part of variables, it is proposed to apply a strengthened Lyapunov function and Lyapunov method. Theorems 2.5.1–2.5.4 are new. To obtain them, the approach to the analysis of the boundedness of solutions of systems of ordinary differential equations described in the monograph of Rumiantsev and Oziraner [1] was used.

2.6. The sufficient conditions for the different types of the μ-boundedness of motion are given under certain assumptions on the dynamic properties of subsystems and limitations on their connection functions. Here some results from the monographs of Michel and Miller [1] and Yoshizawa [2] and from the article of Mitropolsky and V.A. Martynyuk [1] are used.

2.7. Applied problems on the boundedness of solutions of nonlinear engineering systems have been considered in many publications (see Krylov and Bogolyubov [1], Letov [1], Lefschetz [1], Lurie [1], Stocker [1] and the bibliography therein). We only kept to the analysis of some systems of such kind. The results obtained in this section are new and published for the first time.

Different sufficient conditions for the boundedness of solutions of linear and nonlinear systems of ordinary differential equations are available in the works of Burdina [1], Bourland and Haberman [1], Vinograd [1], Gusarova [1], Demidovich [1], Zubov [3], Liu and Shaw [1], Rozo [1], Yakubovich [1], Yakubovich and Starzhinskii [2], and others.

An extensive bibliography of works where questions of the boundedness of motion are studied is available in the monographs of Česari [1], Reissig, Sansone, and Conti [1], and others.

Chapter 3

Analysis of the Stability of Motion

3.1 Introductory Remarks

The analysis of the stability of solutions of nonlinear weakly connected equations is of interest for a number of physical systems, for example, Toda's chains (see Bourland and Haberman [1] and others), as well as systems of weakly connected oscillators (see Goisa and Martynyuk [1] and others).

The application of methods of nonlinear mechanics provides an opportunity to construct asymptotic solutions of such systems and analyze them.

The purpose of this chapter is to determine new sufficient conditions for the μ-stability (μ-instability) of motion of nonlinear weakly connected systems. Those conditions are based on the ideas of the method of comparison with a scalar or vector Lyapunov function.

In Section 3.2, the objectives of the study are formulated and their connection with the problem of stability under continuous perturbations in its classical statement is discussed (see Duboshin [1], Malkin [1]).

In Section 3.3, the direct Lyapunov method and the vector function are applied to obtain sufficient conditions for the stability of a weakly connected system with respect to two measures under the four types of connection functions:

(A_1) bounded at each point of time,

(A_2) asymptotically vanishing at $t \to +\infty$,

(A_3) bounded in the mean, and

(A_4) developing at $t \to +\infty$.

In Section 3.4, the conditions for the stability of the system (3.2.1) are obtained by application of a perturbed Lyapunov function and a scalar comparison equation.

In Section 3.5, the conditions for μ-stability and μ-instability of the equilibrium state of an individual subsystem interacting with other subsystems are found.

In Section 3.6, the algebraic conditions for the uniform asymptotic μ-stability (in the large) and the exponential μ-stability under type A_1 connections are obtained. Here the conditions for μ-instability and complete μ-instability of the equilibrium state of the system (3.2.1) are given.

In Section 3.7, the μ-polystability of the motion of a weakly connected system consisting of two subsystems is discussed.

In Section 3.8, the conditions for the stability of a longitudinal motion of an aeroplane are given, as well as the conditions for the stability of an indirect control system with small linearity and an unstable free subsystem.

3.2   Statement of the Problem

Consider the equations of perturbed motion of a nonlinear weakly connected system in the form

$$\frac{dx_s}{dt} = f_s(t, x_s) + \mu g_s(t, x_1, \ldots, x_m),$$

$$x_s(t_0) = x_{s0}, \quad s = 1, 2, \ldots, m,$$

(3.2.1)

where $x_s \in R^{n_s}$, $f_s \in C(R_+ \times R^{n_s}, R^{n_s})$, $g_s \in C(R_+ \times R^{n_1} \times \ldots \times R^{n_s}, R^{n_s})$, $\mu \in (0, \mu^*]$ is a small parameter. At $\mu = 0$ the system (3.2.1) reduces to the set of unrelated subsystems

$$\frac{dx_s}{dt} = f_s(t, x_s), \quad x_s(t_0) = x_{s0}, \quad s = 1, 2, \ldots, m.$$

(3.2.2)

Apply the two measures $\rho_s(t, x_s)$ and $\rho_{s0}(t, x_s)$ from the class of functions M:

$$M = \{\rho_s \in C(R_+ \times R^{n_s}, R_+), \quad \inf_{t, x_s} \rho_s(t, x_s) = 0, \quad s = 1, 2, \ldots, m\}.$$

In addition, for the measures

$$\rho(t, x) = \sum_{s=1}^{m} a_s \rho_s(t, x_s)$$

(3.2.3)

and

$$\rho_0(t, x) = \sum_{s=1}^{m} a_s \rho_{s0}(t, x_s), \quad a_s = \text{const},$$

(3.2.4)

it is assumed that the inequality

$$\rho(t, x) \leq \varphi(\rho_0(t, x))$$

(3.2.5)

holds provided that

$$\rho_0(t, x) < \delta, \quad \delta > 0,$$

(3.2.6)

where the function φ belongs to the K-class.

For the estimation of the impact of the vector connection function

$$g(t, x) = (g_1(t, x_1, \ldots, x_m), \ldots, g_m(t, x_1, \ldots, x_m))^{\mathrm{T}}$$

the Euclidean norm of the vector $g(t, x)$ is applied:

$$\|g(t, x)\| = \left(\sum_{s=1}^{m} \|g_s(t, x_1, \ldots, x_m)\| \right)^{1/2}.$$

The connection functions $g_s(t, x_1, \ldots, x_m)$, $s = 1, 2, \ldots, m$, in the system (3.2.1) will be considered under certain assumptions:

A_1. The connection functions $g_s \in C(R_+ \times R^{n_1} \times \ldots \times R^{n_m}, R^{n_s})$ at all $s = 1, 2, \ldots, m$ and $\|g(t, x)\| = \left(\sum_{s=1}^{m} \|g_s(t, x_1, \ldots, x_m)\| \right)^{1/2}$ are bounded uniformly with respect to $t \geq t_0 > 0$.

A_2. The connection functions $g_s \in C(R_+ \times R^{n_1} \times \ldots \times R^{n_m}, R^{n_s})$ at all $s = 1, 2, \ldots, m$ and $\lim\limits_{t \to \infty} \|g(t, x)\| = 0$ unifomly with respect to $x \in R^n$.

A_3. The connection functions $g_s \in C(R_+ \times R^{n_1} \times \ldots \times R^{n_m}, R^{n_s})$ at all $s = 1, 2, \ldots, m$ and there exists an integrable function $\varphi(t)$ for which

$$\|g(t, x)\| < \varphi(t), \qquad \int\limits_{t_0}^{t_0+T} \varphi(s) \, ds \leq \Delta$$

for some $T > 0$ and $\Delta > 0$.

A_4. The connection functions $g_s \in C^{(1,1)}(R_+ \times R^{n_1} \times \ldots \times R^{n_m}, R^{n_s})$ at all $s = 1, 2, \ldots, m$ are bounded together with the partial derivatives $\dfrac{\partial g_s}{\partial t}$, $\dfrac{\partial g_s}{\partial x_j}$, $s = 1, 2, \ldots, m$, $j = 1, 2, \ldots, n_s$, and are such that:

(a) $g_s(t_0, x_1, \ldots, x_m) = 0$ at $t_0 \in R_+$ and $x_1 \neq 0, \ldots, x_m \neq 0$, $s = 1, 2, \ldots, m$;

(b) $g_s(t, x_1, \ldots, x_m) \neq 0$ at $t > t_0$, $s = 1, 2, \ldots, m$.

Definition 3.2.1 The system (3.2.1) is said to be (ρ_0, ρ) μ-stable under small bounded interactions of subsystems if for specified $\varepsilon > 0$ and $t_0 \in R_+$ there exist two numbers δ_1, δ_2 and a value of the parameter $\mu^* \in (0, 1]$ such that as soon as

$$\rho_0(t_0, x_0) < \delta_1 \tag{3.2.7}$$

and

$$\|g(t, x)\| < \delta_2 \quad \text{at} \quad (t, x) \in S(\rho, H), \tag{3.2.8}$$

then $\rho(t, x(t; t_0, x_0, \mu)) < \varepsilon$ at all $t \geq 0$ and $\mu < \mu^*$. Here $S(\rho, H) = \{(t, x) \in R_+ \times R^n \colon \rho(t, x) < H\}$, $n = n_1 + n_2 + \ldots + n_m$, $H = \text{const} > 0$.

Remark 3.2.1 The condition (3.2.8) resembles the one applied in the study of the stability under continuous perturbations (see Malkin [1]). However, in this problem the functions $g(t, x)$ are specified as a part of the system (3.2.1) and $g(t, 0) = 0$ at all $t > 0$, and this, as is known, is not assumed in the problem of the stability under continuous perturbations.

3.3 Stability with Respect to Two Measures

Now connect the auxiliary functions $v_s \in C(R_+ \times R^{n_s}, R_+)$, $s = 1, 2, \ldots, m$, $v_s(t, 0) = 0$ at all $t > 0$, with the free subsystems (3.2.2).

The function

$$v(t, x, \beta) = \sum_{s=1}^{m} \beta_s v_s(t, x_s), \quad \beta_s = \text{const} \neq 0, \tag{3.3.1}$$

is assumed to be ρ-positive definite and ρ-decrescent, that is, for this function there exist comparison functions a, b that belong to the K-class and constants Δ_1 and Δ_2 such that

$$\begin{aligned} a(\rho(t, x)) &\leq v(t, x, \beta), & \text{as soon as} \quad \rho(t, x) < \Delta_1, \\ v(t, x, \beta) &\leq b(\rho_0(t, x)), & \text{as soon as} \quad \rho_0(t, x) < \Delta_2 \end{aligned} \tag{3.3.2}$$

respectively.

Let us prove the following statement.

Theorem 3.3.1 *Assume that the equations of perturbed motion (3.2.1) are such that:*

(1) *the state of the subsystems is characterized by the measures $\rho_s(t, x_s)$ and $\rho_{s0}(t, x_s)$ which take on values from the set M;*

(2) *the measure $\rho(t, x) = \sum_{s=1}^{m} \alpha_s \rho_s(t, x_s)$ is uniformly continuous with respect to the measure $\rho_0(t, x) = \sum_{s=1}^{m} \alpha_s \rho_{s0}(t, x_s)$;*

(3) *there exist functions $v_s \in C(R_+ \times R^{n_s}, R_+)$, $s = 1, 2, \ldots, m$, and a function $v(t, x, \beta)$ determined by the formula (3.3.1) is locally Lipschitz with respect to x, ρ-positive definite and ρ_0-decrescent;*

(4) *along solutions of the independent subsystems (3.2.2) the estimate*

$$D^+ v(t, x, \beta)|_{(3.2.2)} \leq -w(\rho_0(t, x)) \tag{3.3.3}$$

holds at all $(t, x) \in S(\rho, H)$, w from the K-class;

(5) *the connection functions* $g_s(t, x_1, \ldots, x_m)$, $s = 1, 2, \ldots, m$, *satisfy the conditions* A_1.

Then the system (3.2.1) is $(\rho_0, \rho)\mu$-stable under small bounded interactions of the subsystems.

Proof The fact that the function $v(t, x, \beta)$ is ρ-positive definite and ρ_0-decrescent implies the existence of constants $\Delta_1 > 0$ and $\Delta_2 > 0$ such that

$$a(\rho(t, x)) \le v(t, x, \beta), \quad \text{if} \quad \rho(t, x) < \Delta_1,$$

and

$$v(t, x, \beta) \le b(\rho_0(t, x)), \quad \text{if} \quad \rho_0(t, x) < \Delta_2.$$

Let $\varepsilon \in (0, \Delta_0)$, where $\Delta_0 = \min(\Delta_1, \Delta_2)$. Choose $\delta_1 \in (0, \Delta_0)$ so that the inequality

$$b(\delta_1) < a(\varepsilon) \quad \text{and} \quad \rho(t, x) < \varepsilon, \tag{3.3.4}$$

should hold as soon as

$$\rho_0(t, x_0) < \delta_1. \tag{3.3.5}$$

The inequality (3.3.4) is possible in view of conditions (1) and (2) of Theorem 3.3.1.

Now construct the function (3.3.1) and calculate $D^+v(t, x, \beta)$ along solutions of the system (3.2.1), taking into account the estimate (3.3.3):

$$D^+v(t, x, \beta)|_{(3.2.1)} \le -w(\rho_0(t, x)) + \mu L \|g(t, x)\| \tag{3.3.6}$$

at all $(t, x) \in S(\rho, H)$, $L > 0$ is the Lipschitz constant for the function $v(t, x, \beta)$.

Choose $\mu^* \in (0, 1]$ and denote $k = \mu/\mu^*$, $\mu < \mu^*$. Taking into account condition (5) of Theorem 3.3.1, choose

$$\delta_2 = \frac{w(\delta_1)}{\mu^* L}.$$

Under the condition (3.3.6) and at the chosen $\delta_2 > 0$ all the conditions of Definition 2.2.1 are satisfied, that is, the system (3.2.1) is (ρ_0, ρ) μ-stable. Let us show this.

Consider the solution $x(t, \mu) = x(t; t_0, x_0, \mu)$ of the system (3.2.1), beginning in the range of values (t_0, x_0), for which $\rho_0(t_0, x_0) < \delta_1$ and $\mu < \mu^*$. Let there exist $t_2 > t_1 > t_0$ such that at $(t, x) \in S(\rho, H) \cap S^c(\rho_0, \delta_1)$

$$\rho_0(t_1, x(t_1, \mu)) = \delta_1, \quad \rho(t_2, s(t_2, \mu)) = H \tag{3.3.7}$$

and

$$\|g(t, x(t, \mu))\| < \delta_2 \quad \text{at all} \quad t \in [t_1, t_2). \tag{3.3.8}$$

From the estimate (3.3.6) under the conditions (3.2.7) and (3.2.8) obtain

$$D^+ v(t, x, \beta)|_{(3.2.1)} < -w(\rho_0(t_1, x(t_1, \mu)))$$
$$+ \frac{\mu}{\mu^*} l \, \frac{w(\rho_0(t_1, x(t_1, \mu)))}{L} = (k-1)w(\delta_1) < 0 \qquad (3.3.9)$$

at all $t_1 < t < t_2$ and $\mu < \mu^*$. Hence obtain the sequence of inequalities

$$a(\rho(t_2, x(t_2, \mu))) \le v(t_2, x(t_2, \mu), \beta)$$
$$\le v(t_1, x(t_1, \mu), \beta) \le b(\rho_0(t_1, x(t_1, \mu)))$$

and taking into account (3.3.4) and (3.3.7), obtain

$$a(\varepsilon) \le v(t_2, x(t_2, \mu), \beta) \le v(t_1, x(t_1, \mu), \beta) < a(\varepsilon).$$

The obtained contradiction invalidates the assumption that there exists $t_2 > t_0$ such that the solution $x(t; t_0, x_0, \mu)$ of the system (3.2.1) at $\mu < \mu^*$ reaches the bound of the domain $S(\rho, H)$ at a point of time $t = t_2$. Hence the system (3.2.1) is (ρ_0, ρ) μ-stable under small bounded interactions of subsystems.

Definition 3.3.1 The system (3.2.1) is said to be asymptotically $(\rho_0, \rho)\mu$-stable under asymptotically decrescent interactions if it is (ρ_0, ρ) μ-stable and for the specified $t_0 \in R_+$ there exist constants $\delta_0 = \delta_0(t_0) > 0$ and $\mu^* \in (0, 1]$ such that $\lim_{t \to \infty} \rho(t, x(t, \mu)) = 0$, as soon as $\rho_0(t_0, x_0) < \delta_0$ and $\mu < \mu^*$.

The following statement contains conditions sufficient for the system (3.2.1) to be asymptotically (ρ_0, ρ) μ-stable under asymptotically decrescent interactions.

Theorem 3.3.2 *Assume that:*

(1) *conditions (1)–(4) of Theorem 3.3.1 are satisfied;*

(2) *the connection functions $g_s(t, x_1, \ldots, x_m)$, $s = 1, 2, \ldots, m$, are asymptotically decrescent, that is, there exists a constant $\sigma > 0$ at which the limit relations*

$$\lim_{t \to \infty} g_s(t, x_1, \ldots, x_m) = 0, \quad s = 1, 2, \ldots, m,$$

are satisfied uniformly with respect to $x_1, \ldots, x_m$ as soon as $\rho(t, x) < \sigma$.

Then the system (3.2.1) is asymptotically (ρ_0, ρ) μ-stable under asymptotically decrescent interactions.

Proof It is clear that under conditions (1) and (2) of Theorem 3.3.2 the system (3.2.1) is (ρ_0, ρ) μ-stable, that is, for $\varepsilon = \min\{\Delta_0, \sigma\}$ there exist constants $\delta_{10} > 0$ and $\delta_{20} > 0$ such that

$$\rho(t, x(t, \mu)) < \sigma_0 \quad \text{at all} \quad t \ge t_0, \quad \mu < \mu^*,$$

as soon as

$$\rho_0(t_0, x_0) < \delta_{10} \quad \text{and} \quad \|g(t, x)\| < \delta_{20}$$

at $(t, x) \in S(\rho, \sigma_0)$.

Further, for $\eta \in (0, \sigma_0)$ choose $\delta_1 = \delta_1(\eta)$ and $\delta_2 = \delta_2(\eta)$ as specified by Definition 3.3.1. According to condition (2) of Theorem 3.3.2 for the quantity

$$\delta_2^* = \min\left\{\delta_2, \frac{w(\delta_1)}{\mu^* L}\right\} \tag{3.3.10}$$

there exists $\tau_1 = \tau_1(t_0, x_0) > 0$ such that

$$\|g(t, x(t, \mu))\| < \delta_2^* \tag{3.3.11}$$

at all $t \geq t_0 + \tau_1$ and $\mu < \mu^*$.

The asymptotic $(\rho_0, \rho)\,\mu$-stability of the system (3.2.1) will be proved if we specify such $\tau = \tau(t_0, x_0) > 0$, that for some $t^* \in [t_0, t_0 + \tau]$ the inequalities

$$\rho_0(t^*, x(t^*, \mu)) < \delta_1$$

and

$$\|g(t, x(t, \mu))\| < \delta_2^*, \quad t \geq t^*$$

will hold. For $\mu^* \in (0, 1]$ such that $k = \mu/\mu^* \leq 1/2$, choose

$$\tau = \frac{4b(\rho_0(t_0 + \tau_1, x(t_0 + \tau_1, \mu)))}{w(\delta_1)} + \tau_1, \quad \mu < \mu^*.$$

Then for the values $t_0 + \tau_1 \leq t \leq t_0 + \tau$ such that $(t, x(t, \mu)) \in S(\rho, \sigma_0) \cap S^c(\rho_0, \delta_1)$, from (3.3.6), (3.3.10), and (3.3.11) obtain the estimate

$$D^+ v(t, x, \beta)|_{(3.2.1)} \leq -\frac{1}{2} w(\delta_1), \quad t_1 + \tau_1 \leq t \leq t_0 + \tau. \tag{3.3.12}$$

Taking into account that the function $v(t, x, \beta)$ is ρ_0-decrescent, from the estimate (3.3.12) obtain the inequality

$$v(t_0+\tau, x(t_0+\tau, \mu), \beta) \leq b(\rho_0(t_0+\tau_1, x(t_0+\tau_1))) - \frac{1}{2} w(\delta_1)(\tau-\tau_1) < 0 \tag{3.3.13}$$

at the chosen τ. But the function $v(t, x, \beta)$ is ρ-positive definite and therefore the obtained contradiction proves the existence of τ, that is, the system (3.2.1) is asymptotically $(\rho_0, \rho)\,\mu$-stable at asymptotic decreases of the connection functions.

Now consider the system (3.2.1) under the conditions of the assumption A_3 on the connection functions $g_s(t, x_1, \ldots, x_m)$, $s = 1, 2, \ldots, m$.

Definition 3.3.2 The system (3.2.1) is said to be (ρ_0, ρ) μ-stable under small in the mean interactions of subsystems if for the specified $\varepsilon > 0$, $t_0 \in R_+$ and $T > 0$ there exist two positive numbers $\delta_1 = \delta_1(\varepsilon)$, $\delta_2 = \delta_2(\varepsilon)$ and a value $\mu^* \in (0, 1]$ such that

$$\rho(t, x(t, \mu)) < \varepsilon \quad \text{at all} \quad t \geq t_0, \quad \mu < \mu^*,$$

as soon as

$$\rho_0(t_0, x_0) < \delta_1, \quad \|g(t, x_1, \ldots, x_m)\| \leq \varphi(t) \quad \text{at} \quad \rho(t, x) \leq \varepsilon, \quad (3.3.14)$$

where

$$\int_t^{t+T} \varphi(s)\, ds < \delta_2.$$

The following statement contains sufficient conditions for (ρ_0, ρ) μ-stability in the sense of Definition 3.3.2.

Theorem 3.3.3 *Assume that:*

(1) *conditions (1) and (2) of Theorem 3.3.1 are satisfied;*

(2) *there exist continuous functions $v_s \in C(R_+ \times R^{n_s}, R_+)$, $s = 1, 2, \ldots, m$, and a function $v(t, x, \beta)$ determined by the formula (3.3.1)*

    (a) *is ρ-positive definite,*

    (b) *is ρ_0-decrescent,*

    (c) *satisfies the Lipschitz condition with respect to x with a constant $L > 0$*
$$|v(t, x, \beta) - v(t, x', \beta)| \leq L\|x - x'\|$$
    at $(t, x), (t, x') \in S(\rho, H)$, $H = \text{const} > 0$;

(3) *there exists a function c from the K-class such that*
$$D^+ v(t, x, \beta)|_{(3.2.2)} \leq -c(v(t, x, \beta)) \quad (3.3.15)$$
at all $(t, x) \in S(\rho, H)$;

(4) *the connection functions $g_s(t, x_1, \ldots, x_m)$, $s = 1, 2, \ldots, m$, satisfy the condition A_3.*

Then the system (3.2.1) is (ρ_0, ρ) μ-stable under small in the mean interactions of subsystems.

Proof Since the function $v(t, x, \mu)$ is ρ-positive definite and ρ-decrescent, for the specified $\varepsilon > 0$, choose $\delta_1 = \delta_1(\varepsilon) > 0$ so that the following inequality would hold:

$$b(\delta_1) < a(\varepsilon). \quad (3.3.16)$$

Let $\rho_0(t_0, x(t_0, \mu)) < \delta_1$ and $m(t, \mu) = v(t, x(t, \mu), \beta)$ at $\mu < \mu^*$. Under the inequality (3.3.16) obtain $m(t_0, \mu) < b(\delta_1) < a(\varepsilon)$ at $\mu < \mu^*$. Show that $m(t, \mu) < a(\varepsilon)$ at all $t \geq t_0$ and $\mu < \mu^*$. Let this statement be incorrect. Then there exists $t_1 > t_0$ such that $m(t_1, \mu) = a(\varepsilon)$ and $m(t, \mu) < a(\varepsilon)$ at $t < t_1$. The inequality

$$a(p(t, x(t, \mu))) \leq v(t, x(t, \mu), \beta) \leq a(\varepsilon)$$

at $t_0 \leq t \leq t_1$ implies the estimate

$$\rho(t, x(t, \mu)) \leq \varepsilon < H, \quad t_0 \leq t \leq t_1. \tag{3.3.17}$$

Let $t_1 - t_0 = T$ and

$$G(u) - G(u_0) = \int_{u_0}^{u} \frac{ds}{c(s)}, \quad G(u) = \int_{0}^{u} \frac{ds}{c(s)}, \quad \text{if} \quad \int_{0}^{u} \frac{ds}{c(s)} < \infty.$$

In the general case

$$G(u) = \int_{\delta}^{u} \frac{ds}{c(s)}$$

for some $\delta > 0$ and $G^{-1}(u)$ is the inverse of the function $G(u)$.

Taking into account conditions (2c) and (3) of Theorem 3.3.3, obtain

$$D^+ v(t, x, \beta)|_{(3.2.1)} \leq -c(v(t, x, \beta)) + \mu L \|g(t, x_1, \ldots, x_m)\| \tag{3.3.18}$$

at $t \in [t_0, t_1]$. To transform the inequality (3.3.18), introduce the notation

$$\lambda(t, \mu) = v(t, x(t, \mu), \beta) - \gamma(t, \mu), \tag{3.3.19}$$

where

$$\gamma(t, \mu) = \mu L \int_{t_0}^{t} \|g(s, x_1(s, \mu), \ldots, x_m(s, \mu))\| \, ds.$$

For the Dini derivative of the function $\lambda(t, \mu)$ obtain the inequality

$$D^+ \lambda(t, \mu) \leq -c(\lambda(t, \mu)), \tag{3.3.20}$$

since the function c is monotone increscent and therefore the inequality $v(t, x, \beta) \leq \lambda(t, \mu)$ implies $c(v(t, x, \beta)) \leq c(\lambda(t, \mu))$.

Applying Bihari's lemma to the inequality (3.3.20), obtain

$$\lambda(t, \mu) \leq G^{-1}[G(v(t_0, x_0, \beta)) - (t - t_0)], \quad t \in [t_0, t_1]. \tag{3.3.21}$$

Now revert to (3.3.21) and note that

$$v(t, x(t, \mu), \beta) \leq G^{-1}[G(v(t_0, x_0, \beta)) - (t - t_0)] + \gamma(t, \mu). \tag{3.3.22}$$

Choose

$$\delta_2 = \delta_2(\varepsilon) < \frac{1}{k\mu^* L}\{a(\varepsilon) - G^{-1}[G(b_1(\delta_1)) - T]\},$$

where $k < 1$.

Taking into account that $\rho(t, x(t, \mu)) \leq \varepsilon$ at all $t_0 \leq t \leq t_0 + T$, $v(t_0, x_0, \beta) < b(\delta_1)$, $\|g(t, x_1, \ldots, x_m)\| \leq \varphi(t)$ and

$$\int\limits_{t_0}^{t_0+T} \varphi(s)\, ds < \delta_2,$$

from the inequality (3.3.22) obtain the following inequality at $t = t_0 + T$:

$$a(\rho(t_0 + T, x(t_0 + T, \mu))) \leq v(t_0 + T, x(t_0 + T, \mu), \beta)$$
$$\leq G^{-1}[G(b_1(\delta_1)) - T] + \mu L \delta_2,$$

or

$$a(\varepsilon) \leq v(t_0 + T, x(t_0 + T, \mu), \beta) + G^{-1}[G(b_1(\delta_1)) - T] + \mu L \delta_2 < a(\varepsilon). \quad (3.3.23)$$

The obtained contradiction proves that $m(t, \mu) < a(\varepsilon)$ at all $t \geq t_0$. Consequently, the system (3.2.1) is (ρ_0, ρ) μ stable at small in the mean connection functions $g_s(t, x_1, \ldots, x_m)$, $s = 1, 2, \ldots, m$.

Now consider the system (3.2.1) at connection functions $g_s(t, x_1, \ldots, x_m)$, $s = 1, 2, \ldots, m$, indicated in the assumption A_4. Such connections are said to be developing (see Martynyuk [6]). In the work the instability of the k-th interacting subsystem in the Lyapunov sense was studied.

Definition 3.3.3 The system (3.2.1) is said to be (ρ_0, ρ) μ-stable under developing connections of subsystems if for specified $\varepsilon > 0$ and $t_0 \in R_+$, $0 < \varepsilon < H$, there exists a number $\delta_1 > 0$ and a value $\mu^* \in (0, 1]$ of the parameter μ such that

$$\rho(t, x(t, \mu)) < \varepsilon \quad \text{at all} \quad t > t_0 \quad \text{and} \quad \mu < \mu^*,$$

as soon as the connection functions $g_s(t, x_1, \ldots, x_m)$, $s = 1, 2, \ldots, m$, satisfy the conditions of the assumption A_4 and $\rho(t_0, x_0) < \delta_1$.

It is necessary to find the conditions for (ρ_0, ρ) μ-stability of the system (3.2.1) under developing connections $g_s(t, x_1, \ldots, x_m)$, $s = 1, 2, \ldots, m$. The solution of this problem is similar to the proof of Theorem 2.2.1 by using the derivative auxiliary function $v(t, x, \beta)$ of an order higher than the first one.

Theorem 3.3.4 *Assume that:*

(1) *conditions (1) and (2) of Theorem 3.3.1 are satisfied;*

(2) *for subsystems (3.2.2) there exist functions $v_s \in C^{(2,2)}(R_+ \times R^{n_s}, R_+)$, $s = 1, 2, \ldots, m$, and the function $v(t, x, \beta)$, determined by the formula (3.3.1) is ρ-positive definite and ρ_0-decrescent;*

(3) *in the domain* $(t, x) \in S(\rho, H)$ *at* $t = t_0$

$$\frac{\partial v}{\partial t} + (\operatorname{grad} v(t, x, \beta))^{\mathrm{T}} f(t, x) \leq 0,$$

where $f(t, x) = (f_1(t, x_1), \ldots, f_m(t, x_m))^{\mathrm{T}}$ *and outside an arbitrarily small neighborhood* $S(\rho, H)$ *at* $t > t_0$

$$\frac{\partial \dot{v}}{\partial t} + (\operatorname{grad} \dot{v})^{\mathrm{T}} (f(t, x) + \mu g(t, x_1, \ldots, x_m)) \leq 0;$$

(4) *in the domain* $(t, x) \in S(\rho, H)$ *there exist constants* $M > 0$ *and* $N > 0$ *such that*

$$\left\| \frac{\partial v}{\partial x} \right\| \leq M, \quad \left\| \frac{\partial \dot{v}}{\partial x} \right\| \leq N$$

and the connection functions $g_s(t, x_1, \ldots, x_m)$, $s = 1, 2, \ldots, m$, *satisfy the conditions of the assumption* A_4.

Then the system (3.2.1) is (ρ_0, ρ) μ-*stable under developing connections* $g_s(t, x_1, \ldots, x_m)$, $s = 1, 2, \ldots, m$.

Proof For the specified $\varepsilon > 0$ and $t_0 \in R_1$ choose $\delta_1 > 0$ like it was set out in the proof of Theorem 3.3.1. Assume that the inequalities (3.3.4) and (3.3.5) hold. From the Lyapunov relation

$$\dot{v}(t, x, \beta) = \dot{v}(t, x, \beta)|_{t=t_0} + \int_{t_0}^{t} \ddot{v}(s, x(s), \beta) \, ds \qquad (3.3.24)$$

under condition (3) of Theorem 3.3.4 obtain

$$[\dot{v}(t, x(t, \mu), \beta)]|_{(3.2.1)} = [\dot{v}(t, x(t, m), \beta)]_{(3.2.1)}|_{t=t_0}$$

$$+ \int_{t_0}^{t} \left[\frac{\partial \dot{v}(t, x, \beta)}{\partial t^2} + (\operatorname{grad} \dot{v}(t, x, \beta))^{\mathrm{T}} [f(t, x) + \mu g(t, x_1, \ldots, x_m)) \right] dt.$$

Since $g_s(t, x_1, \ldots, x_m) = 0$ at $t = t_0$, $s = 1, 2, \ldots, m$, then

$$[\dot{v}(t, x(t, \mu), \beta)]|_{(3.2.1)}|_{t=t_0} = \frac{\partial v}{\partial t} + (\operatorname{grad} v(t, x, \beta))^{\mathrm{T}} f(t, x).$$

Therefore,

$$\dot{v}(t, x(t, \mu), \beta)|_{(3.2.1)}|_{t=t_0} \leq 0 \quad \text{at} \quad (t, x) \in S(\rho, H). \qquad (3.3.25)$$

Hence

$$v(t, x(t, \mu), \beta) \leq v(t_0, x_0, \beta) \quad \text{at all} \quad t \geq t_0. \qquad (3.3.26)$$

Show that if $\rho(t_0, x_0) < \delta_1$, then

$$\rho(t, x(t, \mu)) < \varepsilon, \quad t \geq t_0 \quad \text{at} \quad (t, x) \in S(\rho, H) \tag{3.3.27}$$

and $\mu < \mu^*$. Let this not be so, then there should exist a motion of the system (3.2.1) with the initial values (t_0, x_0): $\rho(t_0, x_0) < \delta_1$ and points of time $t_2 > t_1 > t_0$ such that

$$\rho_0(t_1, x(t_1, \mu)) = \delta_1, \quad \rho(t_2, x(t_2, \mu)) = \varepsilon$$

and

$$\rho(t, x(t, \mu)) \in S(\rho, \varepsilon) \cap S^c(\rho, \delta_1) \tag{3.3.28}$$

at $t \in [t_1, t_2)$. From the relations (3.3.26) and (3.3.4) find

$$a(\varepsilon) \leq v(t_2, x(t_2, \mu), \beta) \leq v(t_1, x(t_1, \mu), \beta) < b(\delta_1) < a(\varepsilon).$$

The obtained contradiction proves that (3.3.27) holds at all $t \geq t_0$, that is, the system (3.2.1) is (ρ_0, ρ) μ-stable under developing connections.

In Theorems 3.3.1–3.3.4 the connection functions $g_s(t, x_1, \ldots, x_m)$, $s = 1, 2, \ldots, m$, were treated as a factor destabilizing the motion of the system (3.2.1) under certain limitations on the dynamic properties of the subsystems (3.2.2) and the connection function the (ρ_0, ρ) μ-stability of the system (3.2.1) may occur due to the fact that connection functions stabilize the motion of the system (3.2.1). This situation is reflected in the following statement.

Theorem 3.3.5 *Assume that:*

(1) *conditions (1) and (2) of Theorem 3.3.1 are satisfied;*

(2) *there exist functions $v_s \in C^{(1,1)}(R_+ \times R^{n_s}, R_+)$, $s = 1, 2, \ldots, m$, and a function $v(t, x, \beta)$, determined by the formula (3.3.1), ρ-positive definite and ρ_0-decrescent;*

(3) *at $(t, x) \in S(\rho, H)$ the following inequality holds:*

$$\sum_{s=1}^{m} \alpha_s \left[\frac{\partial v_s}{\partial t} + (\mathrm{grad}\, v_s(t, x_s))^\mathrm{T} f_s(t, x_s) \right] \leq 0, \quad s = 1, 2, \ldots, m;$$

(4) *the connection functions $g_s(t, x_1, \ldots, x_m)$, $s = 1, 2, \ldots, m$, are such that there exist integrable functions $l_1(t), \ldots, l_m(t)$ for which*

$$\sum_{s=1}^{m} \alpha_s \left[(\mathrm{grad}\, v_s(t, x_s))^\mathrm{T} g_s(t, x_1, \ldots, x_m) \right]$$

$$\leq (l_1(t) + l_2(t) + \cdots + l_m(t)) v(t, x, \beta)$$

and

$$\exp \left[\mu \int_{t_1}^{t_2} (l_1(s) + l_2(s) + \cdots + l_m(s))\, ds \right] \leq N(\mu),$$

$N(\mu) > 0$ *at all $\mu < \mu^*$.*

Then the system (3.2.1) is uniformly (ρ_0, ρ) μ-stable.

Proof For the measures $\rho_0(t, x)$ and $\rho(t, x)$ determined according to condition (1) of Theorem 3.3.5 and the function $v(t, x, \alpha) = \sum_{s=1}^{m} \alpha_s v_s(t, x_s)$ there exist functions a, b from the K-class and constants Δ_1 and $\Delta_2 > 0$ such that

$$v(t, x, \alpha) \leq b(\rho_0(t, x)) \quad \text{at} \quad \rho_0(t, x) < \Delta_2$$

and

$$a(\rho(t, x)) \leq v(t, x, \alpha) \quad \text{at} \quad \rho(t, x) < \Delta_1,$$

where $\Delta_2 \in (0, H)$.

For $\varepsilon \in (0, \Delta_2)$ by choosing $\delta_1 \in (0, \Delta_1)$ secure the satisfaction of the inequality

$$N\delta(\delta_1) < a(\varepsilon) \quad \text{if} \quad \rho(t, x) < \varepsilon \quad \text{and} \quad \rho_0(t, x) < \delta_1. \tag{3.3.29}$$

Let $t_0 \in R_+$ and let $x(t, t_0, x_0, \mu)$ be a solution of the system (3.2.1) with the initial conditions (t_0, x_0) for which

$$\rho_0(t_0, x_0) < \delta_1. \tag{3.3.30}$$

Along this solution, according to conditions (3) and (4) of Theorem 2.2.5 obtain

$$\left.\frac{dv(t, x, \alpha)}{dt}\right|_{(3.2.1)} \leq \sum_{s=1}^{m} \alpha_s \left[\frac{\partial v_s}{\partial t} + (\operatorname{grad} v_s)^{\mathrm{T}} f_s(t, x)\right]$$

$$+ \mu \sum_{s=1}^{m} \alpha_s (\operatorname{grad} v_s)^{\mathrm{T}} g_s(t, x_1, \ldots, x_m) \tag{3.3.31}$$

$$\leq \mu(l_1(t) + \cdots + l_m(t)) v(t, x, \alpha)$$

$$\forall (t, x) \in S(\rho, H) \quad \text{and} \quad \mu < \mu^* \in M.$$

Show that under the conditions (3.3.30) and (3.3.31) the system (3.2.1) is uniformly $(\rho_0, \rho)\mu$-stable. Let this not be so, that is, at (3.3.30) for the solution $x(t, \mu)$ there exist values of time $t_2 > t_1 > t_0$ such that $\rho_0(t_1, x(t_1, \mu)) \leq \delta_1$ and $\rho(t_2, x(t_2, \mu)) = \varepsilon$ and $\rho(t, x(t, \mu)) \in S(\rho, \varepsilon) \cap S^c(\rho_0, \delta_1)$ at all $t \in [t_1, t_2)$. From the inequality (3.3.31) under (3.3.29) obtain

$$a(\varepsilon) \leq v(t_2, x(t_2, \mu), \beta) \leq v(t_1, x(t_1, \mu), \beta) \exp\left[\mu \int_{t_1}^{t_2} (l_1(s) + \cdots\right.$$

$$\left. \cdots + l_m(s)) \, ds\right] \leq N(\mu) b(\delta_1) < a(\varepsilon) \quad \text{at all} \quad \mu < \mu^*.$$

The obtained contradiction proves Theorem 3.3.5.

Remark 3.3.1 Condition (3) of Theorem 3.3.5 is impossible within the limits of the Malkin [1] theorem of stability under continuous perturbations.

3.4 Equistability Via Scalar Comparison Equations

Continue the study of the stability of the nonlinear system (3.2.1) under some additional assumptions.

Let the system (3.2.1) be defined in the domain $R_+ \times D$, $D \subseteq R^n$, and have the unique equilibrium state $x_1 = x_2 = \ldots = x_m = 0$, i.e. $f_s(t,0) = 0$ and $g_s(t,0,\ldots,0) = 0$ at all $s = 1, 2, \ldots, m$. Choose the measures (3.3.3) and (3.2.4) in the form

$$\rho(t,x) = \|x\| = \left(\sum_{s=1}^{m} \|x_s\|^2 \right)^{1/2},$$

$$\rho_0(t,x) = \|x_0\| = \left(\sum_{s=1}^{m} \|x_{s0}\|^2 \right)^{1/2},$$

where $\|\cdot\|$ is the Euclidean norm of the vector x.

Definition 3.4.1 The state of equilibrium $(x = 0) \in R^{n_1} \times \ldots \ldots \times R^{n_m}$ of the system (3.2.1) is equistable, if for the specified $t_0 \in R_+$ and $\varepsilon > 0$ one can find $\delta(t_0, \varepsilon) > 0$ and $\mu^*(\varepsilon) < 1$ such that

$$\|x(t,\mu)\| < \varepsilon \quad \text{at all} \quad t \geq t_0,$$

as soon as $\|x_0\| < \delta(t_0, \varepsilon)$ and $\mu < \mu^*(\varepsilon)$.

Remark 3.4.1 The term "equi" emphasizes the dependence of the property of stability of solutions of the system (3.2.1) on the parameter μ (the condition $\mu < \mu^*(\varepsilon)$).

Similarly to the study of the μ-boundedness, here the dynamic behavior of the subsystems (3.2.2) is characterized by the functions $v_s \in C(R_+ \times D_s, R_+)$ assumed to be locally Lipschitz with respect to $x_s \in D_s$, $D_s \subseteq R^{n_s}$, $s = 1, 2, \ldots, m$.

The impact of the connection functions $\mu g_s(t, x_1, \ldots, x_m)$, $s = 1, 2, \ldots, m$, in the system (3.2.1) upon its state is characterized by the functions $w_s(t, x, \mu)$ defined in the domain $R_+ \times D \cap S^c(\eta) \times M$ at some $\eta > 0$, $S(\eta) = \{x \in R^n : \|x\| < \eta\}$.

Using the functions $v_s(t, x_s)$ and $w_s(t, x, \mu)$, $s = 1, 2, \ldots, m$, we construct the scalar functions

$$v_0(t,x,a) = a^{\mathrm{T}} v(t,x), \quad a \in R_+^m,$$

where $v(t,x) = (v_1(t,x), \ldots, v_m(t, x_m))^{\mathrm{T}}$, and

$$w_0(t,x,\mu,\beta) = \beta^{\mathrm{T}} w(t,x,\mu), \quad \beta \in R^m,$$

where $w(t,x,\mu) = (w_1(t,x,\mu), \ldots, w_m(t,x,\mu))^{\mathrm{T}}$, which will be applied for the determination of the conditions for the μ-stability of the state $x = 0$ of the system (3.2.1).

Theorem 3.4.1 *Assume that the system of equations of perturbed motion (3.2.1) is such that:*

(1) *for the subsystems (3.2.2) there exist functions $v_s \in C(R_+ \times R^{n_s}, R_+)$, $v_s(t, x_s) \geq 0$ at all $s = 1, 2, \ldots, m$, $v_s(t, 0) = 0$ at all $t \in R_+$, and some vector $a \in R_+^m$, $a > 0$, such that*

   (a) *$a(\|x\|) \leq v_0(t, x, a) \leq b(\|x\|)$ at all $(t, x) \in R_+ \times S(H)$, where the functions a, b belong to Hahn's K-class,*

   (b) *$D^+ v_0(t, x, a)|_{(3.2.1)} \leq g_0(t, v(t, x, a), \mu)$ at all $(t, x) \in R_+ \times S(H)$, where $g_0 \in C(R_+ \times R_+ \times M, R)$, $g_0(t, 0, \mu) = 0$ at all $t \in R_+$;*

(2) *for any $\eta > 0$ there exist functions $w_s(t, x, \mu)$ estimating the impact of the connection functions, such that $w_0(t, x, \mu, \beta) \in C(R_+ \times S(H) \cap S^c(\eta) \times M \times R^m, R)$ and*

   (a) *there exists a nondecrescent function $c(\mu)$, $\lim_{\mu \to 0} c(\mu) = 0$, such that*
   $$|w_0(t, x, \mu, \beta)| < c(\mu) \quad \text{at all } t \in R \text{ and } \eta \leq \|x\| < \varepsilon \leq H,$$

   (b) *at all $(t, x, \mu) \in R_+ \times S(H) \cap S^o(\eta) \times M$ the inequality*
   $$D^+ v_0(t, x, a)|_{(3.2.1)} + D^+ w_0(t, x, \mu, \beta)|_{(3.2.1)}$$
   $$\leq g(t, v_0(t, x, a) + w_0(t, x, \mu, \beta), \mu),$$

   is satisfied where $g(t, 0, \mu) = 0$ at all $t \in R_+$, $\mu \in M^0 \subset M$;

(3) *the zero solution of the scalar equation*
   $$\frac{du}{dt} = g_0(t, u, \mu), \quad u(t_0) = u_0 \geq 0, \tag{3.4.1}$$

   is μ-stable;

(4) *the zero solution of the scalar equation*
   $$\frac{dw}{dt} = g(t, w, \mu), \quad w(t_0) = w_0 \geq 0, \tag{3.4.2}$$

   is uniformly μ-stable.

Then the state of equilibrium $x = 0$ of the system (3.2.1) is equistable.

Proof Let $t_0 \in R_+$ and $0 < \varepsilon < H$ be specified. Under conditions (2) and (4) of Theorem 3.4.1 for the function $a(\varepsilon) > 0$ at any $t_0 \in R_+$ one can choose $\delta_0 = \delta_0(\varepsilon) > 0$ and $\mu_1 \in M_1 \subset M$ so that

$$w(t; t_0, w_0, \mu) < a(\varepsilon) \quad \text{at all} \quad t \geq t_0,$$

as soon as

$$w_0 = \sum_{s=1}^{m} \alpha_s v_s(t_0, x_{s0}) + \sum_{s=1}^{m} |\beta_s| \|w_s(t_0, x_0, \mu)\| < \delta_0$$

and $\mu < \mu_1$.

Since the function b belongs to the K-class and is monotone increscent, then for a fixed $\delta_0 > 0$ one can choose $\delta_1 = \delta_1(\varepsilon) > 0$ so that

$$b(\delta_1) < \frac{1}{2}\delta_0(\varepsilon) \quad \text{at} \quad 0 < \varepsilon < H.$$

According to condition (3) of Theorem 3.4.1, the zero solution of the equation (3.4.1) is μ-stable. Therefore, at fixed $\frac{1}{2}\delta_0(\varepsilon) > 0$ an $t_0 \in R_+$ one can choose values $\delta_2 = \delta_2(t_0, \varepsilon) > 0$ and $\mu_2 \in M_2 \subset M$ so that

$$u(t; t_0, u_0, \mu) < \frac{1}{2}\delta_0 \quad \text{at all} \quad t \geq t_0 \qquad (3.4.3)$$

provided that $\mu < \mu_2$ and

$$0 \leq u_0 < \delta_2. \qquad (3.4.4)$$

Note that the inequality (3.4.3) is satisfied for any solution of the equation (3.4.1) with the initial conditions (3.4.4), including the maximum solution, that is, $u^+(t; t_0, u_0, \mu) < \frac{1}{2}\delta_0$ at all $t \geq t_0$.

Let

$$u_0 = \sum_{s=1}^{m} \alpha_s v_s(t_0, x_{s0}), \quad \alpha_s = \text{const} > 0.$$

According to condition (1) of Theorem 3.4.1, the functions $v_s(t, x_s)$, $s = 1, 2, \ldots, m$, are continuous, nonnegative, and vanishing at $x_s = 0$, $s = 1, 2, \ldots, m$. Therefore, for the specified $\delta_2 > 0$ one can choose a value of $\delta_3 > 0$ so that the inequalities

$$\|x_0\| < \delta_3 \quad \text{and} \quad \sum_{s=1}^{m} \alpha_s v_s(t_0, x_{s0}) < \delta_2$$

will be satisfied simultaneously.

Now choose $\delta = \min(\delta_3, \delta_1)$ and show that if $\|x_0\| < \delta$, $\|x_0\| = \left(\sum_{s=1}^{m} \|x_{s0}\|^2\right)^{1/2}$, then the solution $x(t, \mu)$ of the system (3.2.1) will satisfy the estimate

$$\|x(t, \mu)\| < \varepsilon \quad \text{at all} \quad t \geq t_0 \qquad (3.4.5)$$

and $\mu < \mu^*$, where $\mu^* \in M$, that is, it will be equistable in the sense of Definition 3.4.1.

Let $\mu_3 = c^{-1}\left(\frac{1}{2}\delta_0\right)$. Then, according to condition (2a) of Theorem 3.3.1, obtain the estimate

$$\sum_{s=1}^{m} |\beta_s| |w_s(t_0, x_0, \mu)| < c\left(c^{-1}\left(\frac{1}{2}\delta_0\right)\right) = \frac{1}{2}\delta_0 \quad \text{at} \quad \eta \leq \|x\| < \varepsilon. \quad (3.4.6)$$

Let the motion $x(t, \mu)$ of the system begin in the point (t_0, x_0) for which $t_0 \in R_+$ and $\|x_0\| < \delta$, and the inequality (3.4.5) does not hold at all $t \geq t_0$. Since the motion is continuous, for a solution $x(t, \mu)$ there should exist values $t_1, t_2 > t_0$ such that:

(A) $x(t_1; t_0, x_0, \mu) \in \partial S(\delta_1)$;

(B) $x(t_2; t_0, x_0, \mu) \in \partial S(\varepsilon)$;

(C) $x(t; t_0, x_0, \mu) \in \overline{S(\varepsilon) \cap S(\delta_1)}, \quad t \in [t_1, t_2]$.

Let in condition (2) of Theorem 3.4.1 the quantity $\eta = \delta_1$. Then conditions (2a) and (2b) of Theorem 3.4.1 for the function

$$m(t, \mu) = \sum_{s=1}^{m} \alpha_s v_s(t, x_s) + \sum_{s=1}^{m} \beta_s w_s(t, x, \mu), \quad t \in [t_1, t_2],$$

result in the differential inequality

$$D^+ m(t, \mu) \leq g(t, m(t, \mu), \mu), \quad t \in [t_1, t_2]. \quad (3.4.7)$$

From the inequality (3.4.7) and the equality (3.4.2) according to Theorem 1.2.10 is obtained the estimate

$$m(t_2, \mu) \leq w^+(t_2, t_1, m(t_1, \mu), \mu),$$

where $w^+(t_2, \cdot)$ is the maximum solution of the comparison equation (3.4.2) at the initial values (t_1, w_0). Along with the inequality (3.4.7), for the function $v_0(t, x(t, \mu), \alpha)$ we obtain the estimate

$$v_0(t_1, x(t_1, \mu), \alpha) \leq u^+(t_1, t_0, v_0(t_0, x_0, \alpha)),$$

and, according to the condition (3.4.3), obtain

$$v_0(t_1, x(t_1, \mu), \alpha) < \frac{1}{2}\delta_0. \quad (3.4.8)$$

The condition (3.4.8) is the condition for the applicability of Theorem 1.2.10 to the comparison equation (3.4.1).

Taking into account the estimates (3.4.6) and (1a) from Theorem 3.4.1, for the value $t = t_2$ obtain

$$a(\varepsilon) + \sum_{s=1}^{m} |\beta_s| |w_s(t_2, x, \mu)| < w^+(t_2; t_1, w_0, \mu) < a(\varepsilon)$$

or $a(\varepsilon) + \dfrac{1}{2}\delta_0 < a(\varepsilon)$ at $\mu < \mu^*$, $\mu^* = \min\{\mu_1, \mu_2, \mu_3\}$.

The obtained contradiction proves that the assumption of the existence of the value $t_2 > t_0$ for which the solution $x(t, \mu)$ reaches the boundary of the domain $S(\varepsilon)$, that is, the inclusion (B) holds, is incorrect. Thus, $\|x(t, \mu)\| < \varepsilon$ at all $t \geq t_0$ and $\mu < \mu^*$.

Theorem 3.4.1 is proved.

Corollary 3.4.1 If in the conditions of Theorem 3.4.1 the majorizing function $g_0(t, v, \mu) \equiv 0$ and all the remaining conditions of Theorem 3.4.1 are satisfied, then the state of equilibrium $x = 0$ of the system (3.2.1) is equistable.

Corollary 3.4.2 If in the conditions of Theorem 3.4.1 the majorizing function $g(t, v, \mu) \equiv 0$ and all the remaining conditions of Theorem 3.4.1 are satisfied, then the state of equilibrium $x = 0$ of the system (3.2.1) is equistable.

Corollary 3.4.3 If in condition (1b)

$$D^+ v_0(t, x, a)\big|_{(3.2.2)} \leq 0$$

and all the remaining conditions of Theorem 3.4.1 are satisfied, then the state of equilibrium $x = 0$ of the system (3.2.1) is equistable.

Corollary 3.4.4 If in condition (2b)

$$D^+ v_0(t, x, a)\big|_{(3.2.2)} + D^+ w_0(t, x, \mu, \beta)\big|_{(3.2.2)} \leq 0$$

and all the remaining conditions of Theorem 3.4.1 are satisfied, then the state of equilibrium $x = 0$ of the system (3.2.1) is equistable.

Note that condition (1a) and the conditions of Corollary 2.1.3 at $\mu = 0$ are sufficient for the uniform stability of the state $x = 0$ of the subsystems (3.2.2).

3.5 Dynamic Behavior of an Individual Subsystem

The study of the dynamics of an interacting subsystem in the set of systems (3.2.1) is of certain interest, since the subsystems may be unstable or, on the contrary, strongly (e.g., exponentially) stable in themselves (i.e., when isolated).

The purpose of this section is the formulation of conditions sufficient for the μ-stability or μ-instability of the k-th subsystem from the set (3.2.1).

Consider the k-th interacting subsystem of the system (3.2.1)

$$\frac{dx_k}{dt} = f_k(t, x_k) + \mu g_k(t, x_1, \ldots, x_m), \tag{3.5.1}$$

where $x_k(t) \in R^{n_k}$ is the state vector of the subsystem, $f_k \in C(R_+ \times R^{n_k}, R^{n_k})$, $g_k \in C(R_+ \times R^{n_1} \times \ldots \times R^{n_m}, R^{n_k})$. The state of equilibrium of the system (3.5.1) and the free subsystem

$$\frac{dx_k}{dt} = f_k(t, x_k) \qquad (3.5.2)$$

is the state $x_k = 0$ at all $t \in R_+$.

Definition 3.5.1 The state of equilibrium $x_k = 0$ of the k-th interacting subsystem (3.5.1) is said to be μ-stable, if for any $\varepsilon > 0$ and $t_0 \in R_+$ one can find $\delta = \delta(\varepsilon, t_0) > 0$ and $\mu^* > 0$ such that

$$\|x_k(t, t_0, x_0, \mu)\| < \varepsilon \quad \text{at all} \quad t \geq t_0, \qquad (3.5.3)$$

as soon as $\|x_0\| < \delta$ and $\mu < \mu^*$, where $x_0 = (x_{10}^{\mathrm{T}}, \ldots, x_{m0}^{\mathrm{T}})^{\mathrm{T}}$.

Remark 3.5.1 Definition 3.5.1 develops the definition of stability with respect to a part of variables (see Lyapunov [1], Rumiantsev [1], and others) in the sense that, unlike the separation of all the system variables into two groups (in the stability theory with respect to a part of variables), here the k-th vector of state of the system (3.2.1) is considered under various dynamic properties of solutions of the remaining $m - 1$ subsystems.

Theorem 3.5.1 *Assume that the equations of perturbed motion of the k-th interacting subsystem (3.5.1) are such that:*

(1) *there exists a function $v_k \in C(R_+ \times R^{n_k}, R_+)$, $v_k(t, x_k)$ locally Lipschitz with respect to x_k, $v_k(t, 0) = 0$ at all $t \in R_+$, satisfying the inequalities*

    (a) $a(\|x_k\|) \leq v_k(t, x_k) \leq b(\|x_k\|)$ *for all* $(t, x_k) \in R_+ \times S(H_k)$;

    (b) $D^+ v_k(t, x_k)|_{(3.5.2)} \leq g_{0k}(t, v_k(t, x_k), \mu)$, *where* $g_{0k} \in C(R_+ \times R^{n_k} \times M, R)$, $g_{0k}(t, 0, \mu) = 0$ *for all* $t \in R_+$;

(2) *the impact of the connections $g_k(t, x_1, \ldots, x_m)$ is estimated by the function*

$$w_k \in C(R_+ \times S(H_1) \cap S^c(\eta_1) \times \ldots \times S(H_m) \cap S^c(\eta_m), R),$$

$w_k(t, x_1, \ldots, x_m, \mu)$ *is locally Lipschitz with respect to the variables $x_1, \ldots, x_m$, $0 < \eta_s < H_s$, for which*

    (a) *there exists a nondecrescent function $c(\mu)$,* $\lim\limits_{\mu \to \infty} c(\mu) = 0$ *and*

$$|w_k(t, x_1, \ldots, x_m, \mu)| < c(\mu) \text{ for } t \in R_+ \text{ and } 0 \leq \|x\| < \varepsilon \leq H;$$

    (b) *at all* $(t, x, \mu) \in R_+ \times S(H) \cap S^c(\eta) \times M$

$$D^+ v_k(t, x_k)|_{(3.5.2)} + D^+ w_k(t, x_1, \ldots, x_m, \mu)|_{(3.2.1)}$$
$$\leq g_{1k}(t, v_k(t, x_k) + w_k(t, x_1, \ldots, x_m, \mu), \mu),$$

where $g_{1k}(t, 0, \mu) = 0$ at all $t \in R_+$, $\mu \in M^0 \subset M$;

(3) *the zero solution of the equation*

$$\frac{du_k}{dt} = g_{0k}(t, u_k, \mu), \quad u_k(t_0) = u_{k0} \geq 0,$$

is μ-stable;

(4) *the zero solution of the equation*

$$\frac{dv_k}{dt} = g_k(t, u_k, \mu), \quad v_k(t_0) = v_{k0} \geq 0,$$

is uniformly μ-stable.

Then the state of equilibrium $x_k = 0$ of the k-th interacting subsystem (3.5.1) is μ-stable.

The proof of this theorem is similar to that of Theorem 3.4.1 and therefore is not given here.

Note that the influence of the remaining subsystems on the k-th subsystem is estimated by the function w_k, since it contains all variables $x_1, \ldots, x_m$, and by the expression of the derivative $D^+w_k(t, x_1, \ldots, x_m, \mu)|_{(3.2.1)}$ in view of the whole system (3.2.1).

Now consider the subsystem (3.5.1) and determine the conditions for the instability of the equilibrium state $x_k = 0$. Following the works of Chetaev [1] and Martynyuk [15], formulate some definitions.

Definition 3.5.2 The state of equilibrium $x_k = 0$ of the k-th interacting subsystem (3.5.1) is said to be μ-unstable if there exist $\varepsilon > 0$ and $t_0 \in R_+$ such that for any arbitrarily small $\delta > 0$ one can find x_0^*: $\|x_0^*\| < \delta$, $\mu^* \in M$ and $t^* > t_0$ for which $\|x_k(t^*; t_0, x_0^*, \mu)\| \geq \varepsilon$ at $\mu < \mu^*$.

According to the above definition, the μ-instability of the k-th subsystem (3.5.1) will be determined if we only note one path reaching the boundary of the domain $\|x_k\| = H_k$ at arbitrarily small $\|x_0^*\|$.

The subsystem (3.5.1) will be considered in the domain

$$t \geq 0, \quad \|x_k\| \leq H_k,$$
$$\|x_1\| + \ldots + \|x_{k-1}\| + \|x_{k+1}\| + \ldots + \|x_m\| < +\infty, \tag{3.5.4}$$

where $H_k = \text{const} > 0$.

For the subsystem (3.5.2) we construct a function $v_k(t, x_k)$ and give the following definitions.

Definition 3.5.3 A set of points (t, x_k) from the domain (3.5.4), for which $v_k(t, x_k) > 0$, is called the domain $v_k > 0$.

Definition 3.5.4 The function $\Phi(t, x)$ is called positive definite in the domain $v_k > 0$, if for any $\varepsilon > 0$, however small it may be, there exists $\delta(\varepsilon) > 0$ such that for any point (t, x) from the domain (3.5.4) satisfying the condition $v_k(t, x_k) \geq \varepsilon$ the inequality $\Phi(t, x) \geq \delta$ would hold.

Let $v^- : L \to T_0$, $t \to v^-(t; t_0, v_0, \mu)$ be the minimum solution of the equation

$$\frac{dv}{dt} = g(t, v, \mu), \quad v(t_0) \geq v_0, \tag{3.5.5}$$

passing through the point (t_0, v_0) at all $\mu \in M^0 \subseteq M$.

Theorem 3.5.2 *Assume that the equations of perturbed motion of the k-th interacting subsystem (3.5.1) are such that:*

(1) *there exists a function $v_k(t, x_k)$, locally Lipschitz with respect to x_k, and in the domain (3.5.4) the set of points (t_0, x_{k0}) for which $v_k(t_0, x_{k0}) > 0$;*

(2) *for all $(t, x) \in R_+ \times \{v_k > 0\}$ the following estimates hold:*

    (a) *$v_k(t, x_k) \leq b(\|x_k\|)$, where b belongs to the K-class,*

    (b) *$D^+ v_k(t, x_k)|_{(3.5.2)} \geq 0$;*

(3) *there exists a function $w_k \in C(R_+ \times R^N \times M, R)$, $N = n_1 + \ldots + n_m$, $w_k(t, x_1, \ldots, x_m, \mu)$ locally Lipschitz with respect to $x_1, \ldots, x_m$, such that*

    (a) *$|w_k(t, x_1, \ldots, x_m, \mu)| < \chi(\mu)$, $\lim_{\mu \to 0} \chi(\mu) = 0$,*

    (b) *$D^+ v_k(t, x_k)|_{(3.5.1)} + D^+ w_k(t, x_1, \ldots, x_m, \mu)|_{(3.2.1)} \geq g(t, v_k(t, x_k) + w_k(t, x_1, \ldots, x_m, \mu), \mu)$, where $g \in C(R_+ \times R \times M, R)$, $g(t, 0, \mu) = 0$ at all $t \geq t_0$;*

(4) *the zero solution of the equation (3.5.5) is μ-unstable.*

Then the state of equilibrium $x_k = 0$ of the k-th interacting subsystem (3.5.1) is μ-unstable.

Proof Conditions (1) and (2) of Theorem 3.5.2 implies that the state $x_k = 0$ of the free subsystem (3.5.2) is unstable in the sense of Lyapunov. From condition (4) of Theorem 3.5.2 it follows that for the solution $v(t, t_0, v_0, \mu)$ of the equation (3.5.5) there exist ε^*, μ_1 such that for an arbitrarily small δ^* one can find v_0: $0 \leq v_0 \leq \delta^*$ and $\tau > t_0 \in R$, for which

$$v(\tau, t_0, v_0, \mu) > \varepsilon^* \quad \text{at} \quad t \geq \tau. \tag{3.5.6}$$

For the specified $\delta > 0$ choose ε^* so that

$$\begin{gathered} (\forall v_0 : 0 \leq v_0 \leq \delta^*)(\exists x_0 : \|x_0\| < \delta), \\ v_0 \leq v_k(t_0, x_{k0}) + w_k(t_0, x_{10}, \ldots, x_{m0}, \mu). \end{gathered} \tag{3.5.7}$$

Now choose $\mu_2 = \chi^{-1}\left(\frac{1}{2}\delta^*\right)$ and according to condition (3a) of Theorem 3.5.2 obtain

$$|w_k(t, x_1, \ldots, x_m, \mu)| < \frac{1}{2}\delta^* \quad \text{at} \quad \mu < \mu_2 \in M \tag{3.5.8}$$

at all $(t, x) \in R_+ \times R^n$.

Choose ε mentioned in Definition 3.5.2 so that

$$b(\varepsilon) + \frac{1}{2}\delta^* \le \varepsilon^*. \tag{3.5.9}$$

Let $J(t_0, x_0)$ denote the interval of existence of a solution of the system (3.5.1). Let $v_0 \colon 0 \le v_0 \le \delta^*$ and $\tau \ge t_0$ be fixed so that the inequality (3.5.7) holds. If the vector x_0 is chosen so that $\|x_0\| < \delta$ and $\tau \in J(t_0, x_0)$, then the instability $x(t, \mu)$ is determined, as the solution cannot cease to exist without leaving the domain $S(\varepsilon)$.

Now assume that $\tau \in J(t_0, x_0)$, the vector $x_0 \colon \|x_0\| < \delta$, and the inequality (3.5.7) holds. Show that the motion $x_k(t, t_0, x_{0k}, \mu)$ of the subsystem (3.5.1) at $t = \tau$ does not belong to the domain $S(\varepsilon)$. Let this not be so, that is, $\|x_k(t, t_0, x_{0k}, \mu)\| < \varepsilon$ at $t = \tau$. Assume

$$n(t, \mu) = v_k(t, x_k) + |w_k(t, x_1, \ldots, x_m, \mu)|$$

at $t \in [t_0, \tau]$. According to condition (3b), obtain the differential inequality

$$D^+ n(t, \mu) \ge g(t, n(t, \mu), \mu). \tag{3.5.10}$$

Applying the comparison technique to the inequality (3.5.10) and the equation (3.5.5), obtain

$$v_k(\tau, x_k(\tau, t_0, x_{0k}, \mu)) + |w_k(\tau, x_1(\tau, t_0, x_{10} < \mu), \ldots,$$
$$x_m(\tau, t_0, x_{m0}, \mu), \mu)| \ge v^-(\tau, t_0, v_0, \mu). \tag{3.5.11}$$

Here $v^-(\tau, \cdot)$ is the minimum solution of the equation (3.5.5) with the initial conditions (3.5.7).

Taking into account (3.5.6), (3.5.8), and (3.5.9), from (3.5.11) obtain the sequence of inequalities

$$v_k(\tau, x(\tau, t_0, x_k), \mu)) + \frac{1}{2}\delta^* \ge v^-(\tau, t_0, v_0, \mu) > \varepsilon^* \ge b(\varepsilon) + \frac{1}{2}\delta^*. \tag{3.5.12}$$

The inequality (3.5.12) results in a contradiction. It means that

$$x_k(\tau, t_0, x_{k0}, \mu) \notin \operatorname{int} S(\varepsilon) \quad \text{at} \quad t = \tau.$$

Consequently, choosing $\mu^* = \min(\mu_1, \mu_2)$, find that the state of equilibrium $x_k = 0$ of the subsystem (3.5.1) is μ-unstable.

Theorem 3.5.2 is proved.

Remark 3.5.2 In contrast to the conditions of Theorem 19.1 from the monograph of Martynyuk [6], here the perturbed Lyapunov function is used, and it is with the help of the perturbations $w_k(t, x_1, \ldots, x_m, \mu)$ that the influence of the connection functions $\mu g_k(t, x_1, \ldots, x_m)$ between the subsystem (3.5.1) and the remaining $m - 1$ subsystems is estimated.

3.6 Asymptotic Behavior

To characterize the dynamic properties of the system (3.2.1) and the free subsystems (3.2.2) we will use the vector function and the Euclidean norm of the state vector of the system.

3.6.1 Uniform asymptotic stability

The system (3.2.1) will be considered under the condition (A_1) about the interconnection functions $g_i(t, x_1, \ldots, x_m)$, $i = 1, 2, \ldots, m$.

Assuming $\rho_0(t, x) = \rho(t, x_i) = \|x_i\|$, $i = 1, 2, \ldots, m$, formulate the definition of the uniform asymptotic μ-stability, taking into account Definition 2.2.1.

Definition 3.6.1 The state of equilibrium $x = 0$ of the system (3.2.1) is said to be uniformly asymptotically μ-stable if it is uniformly μ-stable and quasiuniformly asymptotically μ-stable.

Following Yoshizawa [2, p. 28], formulate the following definition for the system (3.2.1)

Definition 3.6.2 The state of equilibrium $x = 0$ of the system (3.2.1) is said to be quasiuniformly asymptotically μ-stable if for a specified $\varepsilon > 0$ there exists $\delta_0(\varepsilon) > 0$, $T(\varepsilon) > 0$ and $\mu^* \in (0, 1]$ such that if $\|x_0\| < \delta_0$ and $\mu < \mu^*$, then $\|x(t, t_0, x_0, \mu)\| < \varepsilon$ at all $t \geq t_0 + T(\varepsilon)$.

Remark 3.6.1 Definition 3.6.2 in the above-mentioned monograph was formulated for the system (3.2.1) at $\mu = 0$ and $s = 1$.

Assumption 3.6.1 There exist:

(1) open connected time-invariant neighborhoods $N_i \subseteq R^{n_i}$, $i = 1, 2, \ldots, m$, of the equilibrium states $x_i = 0$ of the subsystems (3.2.2);

(2) continuously differentiable functions $v_i \colon R \times N_i \to R_+$, comparison functions ψ_{i1}, ψ_{i2}, ψ_{i3} from K-class, constants $\sigma_i \in R$ such that

    (a) $\psi_{i1}(\|x_i\|) \leq v_i(t, x_i) \leq \psi_{i2}(\|x_i\|)$,

    (b) $dv_i(t, x_i)/dt|_{(3.2.2)} \leq \sigma_i \psi_{i3}(\|x_i\|)$ in the range of values $(t, x_i) \in R_+ \times N_i$, $i = 1, 2, \ldots, m$;

(3) constants $a_{ij} = a_{ij}(\mu) \in R$ such that

$$(\operatorname{grad} v_i(t, x_i))^{\mathrm{T}} \mu g(t, x_1, \ldots, x_m)$$
$$\leq [\psi_{i3}(\|x_i\|)]^{1/2} \sum_{j=1}^{m} a_{ij}(\mu) [\psi_{j3}(\|x_j\|)]^{1/2}$$

at all $(t, x_i) \in R \times N_i$.

Theorem 3.6.1 *Let the equations of perturbed motion of the weakly connected system (3.2.1) be such that:*

(1) *all the conditions of Assumption 3.6.1 are satisfied;*

(2) *at specified σ_i, $i = 1, 2, \ldots, m$, there exists a vector $\alpha = (\alpha_1, \ldots, \alpha_m)^{\mathrm{T}} > 0$ and a value $\mu^* \in M$ such that a matrix $S = [s_{ij}(\mu)]$ with the elements*

$$s_{ij}(\mu) = \begin{cases} \alpha_i(\sigma_i + a_{ii}(\mu)), & i = j, \\ \dfrac{1}{2}(\alpha_i a_{ij}(\mu) + \alpha_j a_{ji}(\mu)), & i \neq j, \end{cases}$$

is negative definite at all $\mu < \mu^*$.

Then the state of equilibrium $x = 0$ of the system (3.2.1) is uniformly asymptotically μ-stable.

Proof On the basis of the function $v_i(t, x_i)$, $i = 1, 2, \ldots, m$, construct a function

$$v(t, x, \alpha) = \sum_{i=1}^{m} \alpha_i v_i(t, x_i), \tag{3.6.1}$$

for which, according to condition (2a) from Assumption 3.6.1, the estimates

$$\sum_{i=1}^{m} a_i \psi_{i1}(\|x_i\|) \leq v(t, x, \alpha) \leq \sum_{i=1}^{m} a_i \psi_{i2}(\|x_i\|)$$

hold at all $(t, x) \in R_+ \times N_1 \times \ldots \times N_m$. The fact that the functions ψ_{i1}, ψ_{i2} belong to the K-class implies the existence of functions ψ_1, ψ_2 from the K-class such that

$$\psi_1(\|x\|) \leq \sum_{i=1}^{m} a_i \psi_{i1}(\|x_i\|), \quad \psi_2(\|x\|) \geq \sum_{i=1}^{m} a_i \psi_{i2}(\|x_i\|).$$

Hence

$$\psi_1(\|x\|) \leq v(t, x, \alpha) \leq \psi_2(\|x\|), \tag{3.6.2}$$

and therefore the function $v(t, x, \alpha)$ is positive definite and decrescent.

Taking into account conditions (2b) and (3) from Assumption 3.6.1, obtain

$$
\left. \frac{dv(t,x)}{dt} \right|_{(3.2.1)} = \sum_{i=1}^{m} \left\{ \alpha_i \left[\frac{\partial v_i(t,x_i)}{\partial t} + \left(\frac{\partial v_i(t,x_i)}{\partial x_i} \right)^{\mathrm{T}} f_i(t,x_i) \right] \right.
$$

$$
+ \left. \alpha_i \left[\left(\frac{\partial v_i(t,x_i)}{\partial x_i} \right)^{\mathrm{T}} \mu g_i(t,x_1,\dots,x_m) \right] \right\}
$$

$$
= \sum_{i=1}^{m} \left\{ \alpha_i \left[\frac{\partial v_i(t,x_i)}{\partial t} \right]_{(3.2.2)} \right. \qquad (3.6.3)
$$

$$
+ \left. \alpha_i \left[\left(\frac{\partial v_i(t,x_i)}{\partial x_i} \right)^{\mathrm{T}} \mu g_i(t,x_1,\dots,x_m) \right] \right\}
$$

$$
\leq \sum_{i=1}^{m} \{ \alpha_i \sigma_i \psi_{i3}(\|x_i\|) + \alpha_i [\psi_{i3}(\|x_i\|)]^{1/2} \sum_{j=1}^{m} a_{ij}(\mu) [\psi_{i3}(\|x_i\|)]^{1/2} \}.
$$

Introduce the notations: $w = \left\{ [\psi_{13}(\|x_1\|)]^{\frac{1}{2}}, \dots, [\psi_{m3}(\|x_m\|)]^{\frac{1}{2}} \right\}^{\mathrm{T}}$ and $R = [r_{ij}]$ is a matrix with the elements

$$
r_{ij} = \begin{cases} \alpha_i [\sigma_i + a_{ii}(\mu)], & i = j, \\ \alpha_i a_{ij}(\mu), & i \neq j, (i,j) = 1, 2, \dots, m. \end{cases}
$$

From (3.6.3) obtain

$$
\left. \frac{dv(t,x)}{dt} \right|_{(3.2.1)} \leq w^{\mathrm{T}} R w = w^{\mathrm{T}} \left(\frac{1}{2} [R + R^{\mathrm{T}}] \right) w = w^{\mathrm{T}} S(\mu) w.
$$

According to condition (2) of Theorem 3.6.1, there exists $\mu^* \in M$ such that the matrix $S(\mu)$ at $\mu < \mu^*$ is negative definite. Then $\lambda_M(S(\mu)) < 0$ at $\mu < \mu^*$ and

$$
\left. \frac{dv(t,x)}{dt} \right|_{(3.2.1)} \leq \lambda_M(S(\mu)) w^{\mathrm{T}} w = \lambda_M(S(\mu)) \sum_{i=1}^{m} \psi_{i3}(\|x_i\|).
$$

Since ψ_{i3} belongs to the K-class, there exists a function $\psi_3 \in K$ such that

$$
\psi_3(\|x\|) \geq \sum_{i=1}^{m} \psi_{i3}(\|x_i\|),
$$

that is,

$$
\left. \frac{dv(t,x)}{dt} \right|_{(3.2.1)} \leq \lambda_M(S(\mu)) \psi_3(\|x\|), \quad \lambda_M(S(\mu)) < 0 \quad \text{at} \quad \mu < \mu^*
$$

in the range of values $(t,x) \in R_+ \times N_1 \times \dots \times N_m$.

Hence it follows that the state $x = 0$ of the system (3.2.1) is uniformly asymptotically stable.

Remark 3.6.2 If the constants a_{ij} in condition (3) of Assumption 3.6.1 do not depend on μ, then for the elements s_{ij} of the matrix $S(\mu)$ we obtain the expressions

$$s_{ij} = \begin{cases} \alpha a_i [\sigma_i + \mu a_{ii}], & i = j, \\ \dfrac{1}{2}\mu(\alpha_i a_{ij} + \alpha_j a_{ji}), & i \neq j, \end{cases}$$

and Theorem 3.5.1 survives.

3.6.2 The global uniform asymptotic stability

At first formulate the following assumption.

Assumption 3.6.2 For the subsystems (3.2.2):

(1) conditions (1) and (2) of Assumption 3.6.1 with the functions ψ_{i1}, ψ_{i2} from the KR-class are satisfied;

(2) at specified v_i and ψ_{i3} there exist functions $a_{ij} \colon R_+ \times R^n \to R$ such that

$$\left(\frac{\partial v_i(t, x_i)}{\partial x_i} \right)^{\mathrm{T}} g_i(t, x_1, \ldots, x_m)$$

$$\leq [\psi_{i3}(\|x_i\|)]^{1/2} \sum_{j=1}^{m} a_{ij}(t, x)[\psi_{i3}(\|x_i\|)]^{1/2}$$

at all $(t, x) \in R_+ \times R^n$.

For Theorem 3.6.1 formulate a generalization in the following form.

Theorem 3.6.2 *Let the equations of perturbed motions of the weakly connected system (3.2.1) be such that:*

(1) *all the conditions of Assumption 3.6.2 are satisfied;*

(2) *there exists a vector $\alpha^{\mathrm{T}} = (\alpha_1, \ldots, \alpha_m) > 0$, a constant $\varepsilon > 0$, and a value $\mu^* \in M$ such that the matrix $S(t, x, \mu) + \varepsilon E$ is negative definite at all $(t, x, \mu) \in R_+ \times R^n \times M^*$, $M^* \subset M$; here, the elements $s_{ij}(t, x, \mu)$ of the matrix $S(t, x, \mu)$ are defined by the formula*

$$s_{ij}(t, x, \mu) = \begin{cases} \alpha_i [\sigma_i + \mu a_{ii}(t, x)], & i = j, \\ \dfrac{1}{2}\mu(\alpha_i a_{ij}(t, x) + \alpha_j a_{ji}(t, x)), & i \neq j, \end{cases}$$

where E is a unit $(m \times m)$-matrix.

Then the state of equilibrium $x = 0$ of the system (3.2.1) is globally uniformly asymptotically μ-stable.

Proof The function (3.6.1) is positive definite and decrescent. Let $R(t, x, \mu) = [r_{ij}(t, x, \mu)]$ denote an $(m \times m)$-matrix with the elements

$$r_{ij}(t, x, \mu) = \begin{cases} \alpha_i[\sigma_i + \mu a_{ii}(t, x)], & i = j, \\ \mu \alpha_i a_{ij}(t, x), & i \neq j. \end{cases}$$

Using (3.5.19) and the conditions of Assumption 3.6.2, for $\mu < \mu^*$ obtain the estimate

$$\left. \frac{dv(t, x)}{dt} \right|_{(3.2.1)} \leq w^{\mathrm{T}} R(t, x, \mu) w$$

$$= w^{\mathrm{T}} \left(\frac{1}{2} [R(t, x, \mu) + R^{\mathrm{T}}(t, x, \mu)] \right) w \qquad (3.6.4)$$

$$= w^{\mathrm{T}} S(t, x, \mu) w \leq -\varepsilon w^{\mathrm{T}} w = -\varepsilon \sum_{i=1}^{m} \psi_{i3}(\|x_i\|)$$

at all $(t, x) \in R_+ \times R^n$. Since ψ_{i3} belongs to the K-class, there exists ψ_3 from the K-class, such that

$$\psi_3(\|x_i\|) \leq \sum_{i=1}^{m} \psi_{i3}(\|x_i\|).$$

Therefore, from (3.6.4) obtain

$$\left. \frac{dv(t, x, \alpha)}{dt} \right|_{(3.2.1)} \leq -\varepsilon \psi_3(\|x_i\|) \quad \text{at} \quad \mu < \mu^*.$$

Thus, the equilibrium state $x = 0$ of the system (3.2.1) is globally uniformly asymptotically μ-stable.

3.6.3 Exponential stability

Further we will use the following notions.

Definition 3.6.3 The state of equilibrium $x = 0$ of the system (3.2.1) is called to be exponentially μ-stable if in an open connected neighborhood N of the state $x = 0$ one can find constants $r_1, \ldots, r_m > 0$, $a > 0$ and $\lambda > 0$ such that at $t \geq t_0$

$$\|x_1(t, t_0, x_0, \mu)\|^{2r_1} + \ldots + \|x_m(t, t_0, x_0, \mu)\|^{2r_m} \leq a\|x_0\| \exp[-\lambda(t - t_0)].$$

The constants a and λ may depend on N.

Definition 3.6.4 The comparison functions φ_1, φ_2 from the KR-class have the value of the same order if there exist constants α_i, β_i, $i = 1, 2$, such that

$$\alpha_i^{-1} \varphi_i(r) \leq \varphi_j(r) \leq \beta_i^{-1} \varphi_i(r), \quad i \neq j, \quad i, j = 1, 2.$$

Assumption 3.6.3 There exist:

(1) open time-invariant connected neighborhoods $N_i \subseteq R^{n_i}$, $i = 1, 2, \ldots, m$, of the equilibrium states $x_i = 0$ of the subsystems (3.2.2);

(2) continuously differentiable functions $v_i \colon R_+ \times N_i \to R$, comparison functions φ_{i1}, φ_{i2}, which have the values of the same order, φ_{i2} from the K-class, and constants $\sigma_i \in R$ such that

  (a) $u^{\mathrm{T}} A u \le v(t, x, \alpha) \le u^{\mathrm{T}} B u$, where $u = (\|x_1\|^{r_1}, \ldots \ldots, \|x_m\|^{r_m})^{\mathrm{T}}$, $r_1, \ldots, r_m > 0$, A, B are constant $(m \times m)$-matrices,

  (b) $dv_i(t, x_i)/dt\big|_{(3.2.2)} \le \sigma_i \varphi_{i2}(\|x_i\|)$ in the range of values $(t, x_i) \in R_+ \times N_i$, $i = 1, 2, \ldots, m$;

(3) constants $a_{ij} = a_{ij}(\mu) \in R$ such that

$$(\operatorname{grad} v_i(t, x_i))^{\mathrm{T}} \mu g_i(t, x_1, \ldots, x_m)$$
$$\le [\varphi_{i2}(\|x_i\|)]^{1/2} \sum_{j=1}^{m} a_{ij}(\mu)[\varphi_{i2}(\|x_i\|)]^{1/2}$$

at all $(t, x_i) \in R_+ \times N_i$.

Theorem 3.6.3 *Let the equations of perturbed motion of the system (3.2.1) be such that:*

(1) *all the conditions of Assumption 3.6.3 are satisfied;*

(2) *at specified σ_i, $i = 1, 2, \ldots, m$, there exists an m-vector $\alpha^{\mathrm{T}} = (\alpha, \ldots, \alpha_m) > 0$ and $\mu^* \in M$ such that the matrix $S(\mu) = [s_{ij}(\mu)]$ with the elements*

$$s_{ij}(\mu) = \begin{cases} \alpha_i[\sigma_i + a_{ii}(\mu)], & i = j, \\ \dfrac{1}{2}[\alpha_i a_{ij}(\mu) + \alpha_j a_{ji}(\mu)], & i \ne j, \end{cases}$$

is negative definite at $\mu < \mu^*$;

(3) *the matrices A and B in the estimate (2a) of Assumption 3.6.3 are positive definite.*

Then the equilibrium state $x = 0$ of the system (3.2.1) is exponentially μ-stable.

Proof For the function $v(t, x, a) = \alpha^{\mathrm{T}} v(t, x)$ where $v(t, x) = (v_1(t, x_1), \ldots, v_m(t, x_m))^{\mathrm{T}}$, according to condition (3) of Theorem 3.6.3, obtain

$$\lambda_m(A) u^{\mathrm{T}} u \le v(t, x, \alpha) \le \lambda_M(B) u^{\mathrm{T}} u. \tag{3.6.5}$$

Since φ_{i1} belongs to the K-class, there exists a comparison function φ_1 from the K-class, such that $u^\mathrm{T} u \leq \varphi_1(\|x\|)$ and therefore (3.6.5) takes on the form

$$\lambda_m(A)u^\mathrm{T} u \leq v(t,x,\alpha) \leq \lambda_M(B)\varphi_1(\|x\|). \tag{3.6.6}$$

Using (3.6.3), conditions (2b) and (3) of Theorem 3.6.3, obtain

$$\left.\frac{dv(t,x,\alpha)}{dt}\right|_{(3.2.1)} \leq \lambda_M(S(\mu)) \sum_{i=1}^{m} \varphi_{i2}(\|x_i\|) \tag{3.6.7}$$

at all $(t,x_i) \in R_+ \times N_i$, $i = 1,2,\ldots,m$. Since φ_{i2} belongs to the K-class, there exists φ_2 from the K-class, such that

$$\sum_{i=1}^{m} \varphi_{i2}(\|x_i\|) \leq \varphi_2(\|x\|).$$

Taking into account the last estimate, the inequality (3.6.7) will take on the form

$$\left.\frac{dv(t,x,\alpha)}{dt}\right|_{(3.2.1)} \leq \lambda_M(S(\mu))\varphi_2(\|x\|). \tag{3.6.8}$$

Under the hypothesis of Theorem 3.6.3 there exists $\mu^* \in M$ such that

$$\lambda_M(S(\mu)) < 0 \quad \text{at all} \quad \mu < \mu^*. \tag{3.6.9}$$

Since the functions φ_1 and φ_2 are the values of the same order, there exist constants β_1, β_2 such that

$$\beta_1^{-1}\varphi_1(r) \leq \varphi_2(r) \leq \beta_2^{-1}\varphi_1(r). \tag{3.6.10}$$

Taking into account (3.6.8) and (3.6.9), for (3.6.7) obtain

$$\left.\frac{dv(t,x,\alpha)}{dt}\right|_{(3.2.1)} \leq -\lambda_M(S(\mu))\beta_1^{-1}\lambda_M^{-1}(B)v(t,x,\alpha),$$

whence

$$v(t,x(t),\alpha) \leq v(t_0,x_0,\alpha)\exp\left[\frac{\lambda_M(S(\mu))}{\beta_1\lambda_M(B)}(t-t_0)\right], \quad t \geq t_0. \tag{3.6.11}$$

Taking into account the inequality

$$\|x_1\|^{2r_1} + \ldots + \|x_m\|^{2r_m} \leq (\|x_1\|^{r_1},\ldots,\|x_m\|^{r_m})^\mathrm{T}(\|x_1\|^{r_1},\ldots,\|x_m\|^{r_m}) \tag{3.6.12}$$

and the estimates (3.6.6) and (3.6.11), obtain

$$\|x_1(t,t_0,x_0)\|^{2r_1} + \ldots + \|x_m(t,t_0,x_0)\|^{2r_m}$$

$$\leq \lambda_m^{-1}(A)\lambda_M(B)\varphi_1(\|x\|_0)\exp\left[\frac{\lambda_M(S(\mu))}{\beta_1\lambda_M(B)}(t-t_0)\right], \quad t \geq t_0. \tag{3.6.13}$$

Assuming

$$a = \lambda_m^{-1}(A)\lambda_M(B) \quad \text{and} \quad \lambda = \frac{\lambda_M(S(\mu))}{\beta_1 \lambda_M(B)},$$

from the estimate (3.6.13) obtain the inequality involved in Definition 3.6.3. Theorem 3.6.3 is proved.

Assumption 3.6.4 There exist:

(1) continuously differentiable functions $v_i \colon R_+ \times R^{n_i} \to R$, constant $(m \times m)$-matrices A and B, constants $r_1, \ldots \ldots, r_m > 0$ such that

(a) $u^T A u \leq v(t, x, a) \leq w_1^T B w_1$, where

$$w_1 = (c_{11}\|x_1\|^2, \ldots, c_{1m}\|x_m\|^2)^T,$$
$$u = (\|x_1\|^{r_1}, \ldots, \|x_m\|^{r_m})^T;$$

(b) $dv_i(t, x_i)/dt\big|_{(3.2.2)} \leq \sigma_i \|x_i\|^2$ in the range of values $(t, x_i) \in R_+ \times R^{n_i}$, $i = 1, 2, \ldots, m$;

(2) at specified σ_i, $i = 1, 2, \ldots, m$, there exist constants $a_{ij} \in R$ such that

$$(\operatorname{grad} v_i(t, x))^T g_i(t, x_1, \ldots, x_m) \leq \|x_i\| \sum_{j=1}^{m} a_{ij}\|x_j\|$$

at all $x_i \in R^{n_i}$, $x_j \in R^{n_j}$, $i, j = 1, 2, \ldots, m$.

Theorem 3.6.4 *Let the equations of perturbed motion of the weakly connected system (3.2.1) be such that:*

(1) *all the conditions of Assumption 3.6.4 are satisfied;*

(2) *at specified σ_i, $i = 1, 2, \ldots, m$, there exists $\mu^* \in (0,1]$ such that the matrix $S(\mu) = [s_{ij}(\mu)]$ with the elements*

$$s_{ij}(\mu) = \begin{cases} \alpha_i(\sigma_i + \mu a_{ii}), & i = j, \\ \frac{1}{2}\mu(\alpha_i a_{ij} + \alpha_j a_{ji}), & i \neq j, \end{cases}$$

is negative definite at $\mu < \mu^*$;

(3) *the matrices A and B in the estimate (1a) of Assumption 3.6.4 are positive definite.*

Then the equilibrium state $x = 0$ of the system (3.2.1) is globally exponentially μ-stable.

Proof From the conditions of Assumption 3.6.4 it follows that the comparison functions $\varphi_{1i}, \varphi_{2i}$ have the form

$$\varphi_{i1}(\|x_i\|) = c_{1i}\|x_i\|^2, \quad \varphi_{i2}(\|x_i\|) = \|x_i\|^2, \quad i = 1, 2, \ldots, m.$$

It is clear that those function have the value of the same order. In addition, the function $v(t, x, \alpha)$ is positive definite, decrescent, and radially unbounded.

The function $dv(t, x, \alpha)/dt$ along solutions of the system (3.2.1) is negative definite in view of condition (1) of Theorem 3.6.4. Hence the state $x = 0$ of the system (3.2.1) is globally exponentially μ-stable.

Theorem 3.6.4 is proved.

3.6.4   Instability and full instability

In Section 3.5 the conditions for the μ-instability of an individual subsystem in the system (3.2.1) were determined. Here we will consider the complete system (3.2.1) under the following conditions.

Assumption 3.6.5   There exist:

(1) open connected time-invariant neighborhoods $N_i \subseteq R^{n_i}$ of the equilibrium states $x_i = 0$ of the subsystems (3.2.2);

(2) continuously differentiable functions $v_i \colon R_+ \times N_i \to R_+$, comparison functions $\psi_{i1}, \psi_{i2}, \psi_{i3}$ from the K-class, constants $\delta_{i1}, \delta_{i2}, \sigma \subset R$ such that

   (a) $\delta_{i1}\psi_{i1}(\|x_i\|) \leq v_i(t, x_i) \leq \delta_{i2}\psi_{i2}(\|x_i\|)$,

   (b) $dv_i(t, x_i)/dt\big|_{(3.2.2)} \leq \sigma_i\psi_{i3}(\|x_i\|)$ in the range of values $(t, x_i) \in R_+ \times N_i$, $i = 1, 2, \ldots, m$;

(3) constants $a_{ij} \in R$ and a value $\mu^* \in (0, 1]$ such that

$$\operatorname{grad} v_i(t, x_i)\mu g_i(t, x_1, \ldots, x_m) \leq [\psi_{i3}(\|x_i\|)]^{1/2} \sum_{j=1}^{m} a_{ij}[\psi_{j3}(\|x_j\|)]^{1/2}$$

at all $(t, x_i, x_j) \in R_+ \times N_i \times N_j$ and $\mu < \mu^*$.

Remark 3.6.3 If $\delta_{i1} = \delta_{i2} = -1$, then it is said that the subsystems (3.2.2) have the property C, and if $\delta_{i1} = \delta_{i2} = 1$ they have the property A.

Remark 3.6.4 From conditions (1) and (2) of Assumption 3.6.4 it follows that if the subsystems (3.2.2) have the property C and $\sigma_i < 0$ at all $i = 1, 2, \ldots, m$, then the equilibrium state $x_i = 0$ of the subsystems (3.2.2) is quite unstable, that is, all the subsystems (3.2.2) are unstable in the sense of the Lyapunov definition [1].

If the subsystems (3.2.2) have the property A and $\sigma_i < 0$ at all $i = 1, 2, \ldots, m$, then all the subsystems (3.2.2) are uniformly asymptotically stable.

Let $L = \{1, 2, \ldots, m\}$ be the set of all subsystems in the complex system (3.2.1). Let $N \neq \varnothing$ denote a set of subsystems that have the property C, $N \subset L$.

Theorem 3.6.5 *Let the equations of perturbed motion (3.2.1) be such that:*

(1) *all the conditions of Assumption 3.6.5 are satisfied;*

(2) *at specified σ_i there exists a vector $\alpha^T = (\alpha_1, \ldots \ldots, \alpha_m) > 0$ and a value $\mu^* \in M$ such that the matrix $S(\mu) = [s_{ij}(\mu)]$ with the elements*

$$s_{ij}(\mu) = \begin{cases} \alpha_i[\sigma_i + a_{ii}(\mu)], & i = j, \\ \dfrac{1}{2}[\alpha_i a_{ij}(\mu) + \alpha_j a_{ji}(\mu)], & i \neq j, \end{cases}$$

is negative definite at $\mu < \mu^*$.

Then:

(a) *if $N \neq L$, then the equilibrium state $x = 0$ of the system (3.2.1) is μ-unstable;*

(b) *if $N = L$, then the equilibrium state $x = 0$ of the system (3.2.1) is completely μ-unstable.*

Proof Like in Theorem 3.6.1, for the function

$$v(t, x, \alpha) = \sum_{s=1}^{m} \alpha_s v_s(t, x_s) \tag{3.6.14}$$

its negative definiteness and decrease are determined. Similarly to (3.6.4) obtain the estimate

$$\left. \frac{dv(t, x, \alpha)}{dt} \right|_{(3.2.1)} \le \lambda_M(S(\mu)) \sum_{i=1}^{m} \psi_{i3}(\|x_i\|) \tag{3.6.15}$$

at all $(t, x) \in R_+ \times N_1 \times \ldots \times N_m$, $i \in L$. According to condition (2) of Theorem 3.6.5 at $\mu \le \mu^*$ $\lambda_m(S(\mu)) < 0$, and therefore $dv(t, x, \alpha)/dt\big|_{(3.2.1)}$ is negative definite.

Now consider the set $D = \{(t, x) \in R_+ \times R^n \colon x_i \in B_i(r)$ at $i \in N$ and $x_i = 0$, as soon as $i \notin N\}$. Here $r = \min\limits_i r_i$, for which $i \in N$. For the function (3.6.14) obtain the estimate

$$-\sum_{i \in N} \alpha_i \psi_{i1}(\|x_i\|) \le v(t, x, \alpha) \le -\sum_{i \in N} \alpha_i \psi_{i2}(\|x_i\|).$$

Hence it follows that in any neighborhood of the state $x = 0$ there exists at least one point $x^* \neq 0$ in which $v(t, x^*, \alpha) < 0$ at all $t \in R_+$. In addition, over the set D the function $v(t, x, a)$ is bounded below. Thus, all the conditions of the Lyapunov theorem [1] on instability at $\mu < \mu^*$ are satisfied. This proves statement (a) of the theorem if $N \neq L$. If $N = L$, all the subsystems (3.2.2) are unstable, and the connections do not change that dynamical state, that is, the system (3.2.1) is completely μ-unstable.

3.7   Polystability of Motion

The polystability of motion of dynamic systems is some extension of the concept of stability with respect to a part of variables.

The purpose of this section is the study of the μ-polystability of the system (3.2.1) at $m = 2$.

3.7.1   General problem of polystability

Consider the nonlinear system of equations of perturbed motion

$$\frac{dx_i}{dt} = f_i(t, x_i) + \mu g_i(t, x_1, \ldots, x_m),$$

$$x_i(t_0) = x_{i0}, \quad i = 1, 2, \ldots, m, \tag{3.7.1}$$

where $x_i \in R^{n_i}$, $t \in R_+$, $t_0 \in R_i$, $R_i \subseteq R$, $f_i \colon R_+ \times R^{n_i} \to R^{n_i}$, $g_i \colon R_+ \times R^{n_1} \times R^{n_2} \times \ldots \times R^{n_m} \to R^{n_i}$, and assume that $f_i(t, 0) = g_i(t, 0, \ldots, 0) = 0$ for all $t \in R_+$, $\mu \in (0, 1]$.

Definition 3.7.1 The system (3.7.1) is said to be μ-polystable (on R_+) if and only if its solution $(x = 0) \in R^n$ is μ-stable (on R_+) and μ-attracting (on R_+) with respect to a group of variables $\{x_i\}^{\mathrm{T}}$, $i = 1, 2, \ldots, m$ (with respect to a set of groups of variables $\{x_1^{\mathrm{T}}, \ldots, x_l^{\mathrm{T}}\}$, $l < m$).

Remark 3.7.1   If the μ-polystability of the equilibrium state $x = 0$ is considered with respect to all subvectors $\{x_i\}^{\mathrm{T}}$, $i = 1, 2, \ldots, m$, of the system (3.7.1), then the system (3.7.1) is considered in the domain

$$R_+ \times B_1(\rho) \times B_2(\rho) \times \ldots \times B_i(\rho), \quad B_i(\rho) = \{x_i \colon \|x_i\| < H_i\},$$

$$H_i = \text{const} > 0, \quad i = 1, 2, \ldots, m,$$

or in R^n, as usual.

Remark 3.7.2   If the μ-polystability of the equilibrium state $x = 0$ of the system (3.7.1) is considered with respect to the set of subvectors $(x_1^{\mathrm{T}}, \ldots, x_l^{\mathrm{T}})$, $l < m$, then the system (3.7.1) is considered in the domain

$$B_l(\rho) = \{x_i^{\mathrm{T}} \colon \|(x_1^{\mathrm{T}}, \ldots, x_l^{\mathrm{T}})^{\mathrm{T}}\| < H^*\}, \quad H^* = \text{const} > 0,$$

$$D_l = \{x_k^{\mathrm{T}} \colon 0 < \|(x_{l+1}^{\mathrm{T}}, \ldots, x_m^{\mathrm{T}})^{\mathrm{T}}\| < +\infty\}, \quad k = l+1, \ldots, m.$$

Here the motion $x(t, x_0, \mu)$ of the system (3.7.1) should be defined at all $t \in R_+$, for which $\|(x_1^{\mathrm{T}}, \ldots, x_l^{\mathrm{T}})^{\mathrm{T}}\| < H^*$.

This condition is satisfied in all applied problems (see Rumiantsev and Oziraner [1]), since it means that none of the coordinates of the subvectors $x_k^{\mathrm{T}}(t)$, $k = l+1, \ldots, m$, of the state of the subsystems (3.7.1) reaches infinity in a finite period of time.

3.7.2 Polystability of the system with two subsystems

Assume that for the independent subsystems

$$\frac{dx_i}{dt} = f_i(t, x_i), \quad x_i(t_0) = x_{i0}, \quad i = 1, 2, \tag{3.7.2}$$

the functions $v_i \colon R_+ \times R^{n_i} \to R$, $i = 1, 2$, are constructed, as well as the function

$$v(t, x, \alpha) = \alpha_1 v_1(t, x_1) + \alpha_2 v_2(t, x_2), \quad \alpha_1, \alpha_2 = \text{const.} \tag{3.7.3}$$

See the following definitions.

Definition 3.7.2 The function $v(t, x, a) \colon R_+ \times R^n \times R \to R_+$ is

(1) positive semidefinite on R_+, if there exists a time-invariant neighborhood N of the state $x = 0$, such that

    (a) $v(t, x, \alpha)$ is continuous with respect to $(t, x) \in R_+ \times N$;

    (b) $v(t, x, \alpha)$ is nonnegative on N, that is,

$$v(t, x, \alpha) \geq 0 \quad \forall (t, x) \in R_+ \times N;$$

    (c) $v(t, x, \alpha) = 0$, if $x = 0$ at all $t \in R_+$;

(2) x_i^{T}-positive definite on R_+, if in the domain $R_+ \times B_i(\rho) \times D_j$, $i \neq j$, $i, j = 1, 2$, the following conditions are satisfied:

    (a) $v(t, x, \alpha)$ is continuous with respect to $(t, x) \in R_+ \times B_i(\rho) \times D_j$;

    (b) there exists a function $w(x_i^{\mathrm{T}})$ such that the inequality

$$w(x_i^{\mathrm{T}}) \leq v(t, x, \alpha) \quad \forall (t, x, \alpha) \in R_+ \times B_i(\rho) \times D_j$$

    holds for one of the values $i = 1, 2$;

    (c) $v(t, x_1, x_2, \alpha) = 0$, if $x_1 = 0$, $x_2 \neq 0$, or $x_1 \neq 0$, $x_2 = 0$;

(3) x_i^{T}-decrescent on R_+, $i = 1, 2$, if in the domain $R_+ \times B_i(\rho) \times D_j$, $i \neq j$, $i, j = 1, 2$, condition (2a) is satisfied and there exists a function $\tilde{w}(x_i^{\mathrm{T}})$ such that

$$v(t, x, \alpha) \leq \tilde{w}(x_i^{\mathrm{T}}), \quad i = 1, 2.$$

The system (3.7.1) at $s = 2$ is further considered in the domain $R_+ \times B_1(H_1) \times B_2(H_2)$.

Definition 3.7.3 The system (3.7.1) at $s = 2$ is said to be μ-polystable (on R_+), if its equilibrium state $x = 0$ is

(a) uniformly μ-stable on R_+ with respect to $(x_1^{\mathrm{T}}, x_2^{\mathrm{T}})$,

(b) uniformly asymptotically μ-stable on R_+ with respect to x_2^{T}.

Theorem 3.7.1 *Let the equations of perturbed motion (3.7.1) at $s = 2$ be such that:*

(1) *there exist differentiable functions $v_i(t, x_i)\colon R_+ \times R^{n_i} \to R$, $i = 1, 2$, for which the function $v(t, x, \alpha)$ is*

    (a) *positive definite on $N \subseteq R^n$, $n = n_1 + n_2$,*

    (b) *decrescent on N (on $R_+ \times N$);*

(2) *there exists $\mu^* \in M$ such that the function $dv(t, x, a)/dt\big|_{(3.7.1)}$ is*

    (a) *negative semidefinite on $R_+ \times N$ at $\mu < \mu^*$,*

    (b) *x_2^{T}-negative definite on $R_+ \times N$ at $\mu < \mu^*$.*

Then the system (3.7.1) is μ-polystable in the sense of Definition 3.7.3.

Proof From conditions (1a), (1b), and (2a), it follows that the state $(x = 0) \in R^n$, $n = n_1 + n_2$, is uniformly μ-stable on R_+. If in addition condition (2b) is satisfied, then the uniform asymptotic μ-stability on R_+ occurs with respect to x_2^{T}.

Now consider the system (3.7.1) at $s = 2$ in the domain

$$R_+ \times B_1(H_1) \times D_2, \quad D_2 = \{x_2\colon 0 < \|x_2\| < +\infty\}.$$

Extend Theorem 4.2 from the monograph by Rumiantsev and Oziraner [1] to the systems (3.7.1) at $s = 2$.

Theorem 3.7.2 *Let the equations of perturbed motion (3.7.1) at $s = 2$ be such that:*

(1) *there exist differentiable functions $v_i(t, x_i)\colon R_+ \times R^{n_i} \to R_+$, $i = 1, 2$, for which the function $v(t, x, \alpha)$ is*

    (a) *x_1^{T}-positive definite on $R_+ \times B_1(H_1) \times D_2$,*

    (b) *decrescent on $R_+ \times B_1(H_1) \times D_2$,*

    (c) *x_1^{T}-decrescent on $R_+ \times B_1(H_1) \times D_2$;*

(2) *the function $dv(t, x, \alpha)/dt\big|_{(3.7.1)}$ is*

    (a) *negative semidefinite on $R_+ \times B_1(H_1) \times D_2$,*

    (b) *x_1^{T}-negative definite on $R_+ \times B_1(H_1) \times D_2$,*

    (c) *negative definite on $R_+ \times B_1(H_1) \times D_2$.*

Then, respectively,

(a) *conditions (1a) and (2a) are sufficient for the* $x_1^{\mathrm{T}} - \mu$*-stability of the state* $(x = 0) \in R^n$ *on* R_+;

(b) *conditions (1b) and (2b) are sufficient for the uniform* $x_1^{\mathrm{T}} - \mu$*-stability of the state* $(x = 0) \in R^n$ *on* R_+;

(c) *conditions (1c) and (2c) are sufficient for the asymptotic* $x_1^{\mathrm{T}} - \mu$*-stability of the state* $(x = 0) \in R^n$ *on* R_+.

The proof of this theorem lies in the verification of the satisfaction of the conditions of Theorems 5.1, 5.2, and 6.1 from the above-mentioned monograph under the conditions of Theorem 3.7.2.

In connection with Theorem 3.7.1 the question that has to be answered is what general form the system (3.7.1) should have for its motion to be μ-polystable in the sense of Definition 3.7.3. The answer resides in the following assumption.

Assumption 3.7.1 There exist:

(1) continuously differentiable functions $v_i \colon R_+ \times R^{n_i} \to \to R_+$, $i = 1, 2$, (2×2)-constant matrices A_1 and B_1 and the comparison function ψ from the K-class, such that

    (a) $u^{\mathrm{T}} A_1 u \le v(t, x, a) \le u^{\mathrm{T}} B_1 u$ at all $(t, x) \in R_+ \times B_1(H_1) \times B_2(H_2)$, where $u^{\mathrm{T}} = (\|x_1\|, \|x_2\|)$,

    (b) $\partial v_1/\partial t + (\partial v_1(t, x_1)/\partial x_k)^{\mathrm{T}} f_1(t, x_1) \le 0$, $k = 1, 2, \ldots, n_1$,

    (c) $\partial v_2/\partial t + (\partial v_2(t, x_2)/\partial x_l)^{\mathrm{T}} f_2(t, x_1) \le -\sigma_1 \psi(\|x_2\|)$, $l = 1, 2, \ldots, n_2$;

(2) there exist constants $k_1, k_2 > 0$ such that

$$|\partial v_1(t, x_1)/\partial x_k| < k_1 \quad \text{and} \quad |\partial v_2(t, x_2)/\partial x_j| < k_2,$$

$k - 1, 2, \ldots, n_1$, $j = 1, 2, \ldots, n_2$;

(3) the connection functions of the subsystems $g_i(t, x_1, x_2)$ satisfy the conditions

$$g_1(t, x_1, x_2) = g_1(t, x_2),$$
$$g_2(t, x_1, x_2) \equiv 0,$$

and, in addition, $\|g_1(t, x_2)\| \le \psi(\|x_2\|)$, where ψ belongs to the K-class.

It is easy to show that under all conditions of Assumption 3.7.1 the system

$$\frac{dx_1}{dt} = f_1(t, x_1) + \mu g_1(t, x_2),$$
$$\frac{dx_2}{dt} = f_2(t, x_2) \tag{3.7.4}$$

is μ-polystable in the sense of Definition 3.7.3.

From conditions (1b) and (1c) it follows that the subsystem

$$\frac{dx_1}{dt} = f_1(t, x_1)$$

is neutrally stable and the subsystem

$$\frac{dx_2}{dt} = f_2(t, x_2)$$

is uniformly asymptotically stable. Hence, at $\mu < \mu^*$ the function $\mu g_1(t, x_2)$ has a stabilizing impact in the system (3.7.4).

3.8 Applications

In this section, some typical systems of the theory of automatic control are considered on the basis of the general approach developed in this chapter.

3.8.1 Analysis of longitudinal motion of an aeroplane

The controllable longitudinal motion of an aeroplane may be described by the equations (see Letov [1] and others)

$$\frac{dx_k}{dt} = -\rho_k x_k + \mu\sigma, \quad k = 1, 2, 3, 4,$$

$$\frac{d\sigma}{dt} = -rp\sigma - f(\sigma) + \mu\sum_{k=1}^{4} \beta_k x_k, \tag{3.8.1}$$

where $\rho_k > 0$, $r > 0$, $p > 0$, β_k are constants, μ is a small parameter, $x_k \in R$, $\sigma \in R$, the function $f\colon R \to R$ has the following properties:

(a) f is continuous on R;

(b) $f(\sigma) = 0$, if and only if $\sigma = 0$;

(c) $\sigma f(\sigma) > 0$ at all $\sigma \neq 0$.

The function f with the properties (a)–(c) is called the admissible non-linearity for the system (3.8.1) (see Aizerman and Gantmacher [1]). If the equilibrium state

$$x^{\mathrm{T}} = (x_1, x_2, x_3, x_4, \sigma) = 0 \tag{3.8.2}$$

of the system (3.8.1) is globally asymptotically stable at all admissible non-linearities of f, then the system (3.8.1) is absolutely stable.

At $\mu = 0$ from the system (3.8.1) obtain two isolated subsystems

$$\frac{dx_k}{dt} = -\rho_k x_k, \quad k = 1, 2, 3, 4, \tag{3.8.3}$$

$$\frac{d\sigma}{dt} = -rp\sigma - f(\sigma). \tag{3.8.4}$$

Represent the system (3.8.1) in the form

$$\frac{dz_i}{dt} = f_i(z_i) + \mu \sum_{l=1, j \neq i}^{2} C_{ij} z_j, \quad i = 1, 2, \tag{3.8.5}$$

where the matrices C_{12}^{T} and C_{21} are such

$$C_{12}^{\mathrm{T}} = [1, 1, 1, 1] \quad \text{and} \quad C_{21} = [\beta_1, \beta_2, \beta_3, \beta_4]. \tag{3.8.6}$$

Now assume that

$$\rho_1 \leq \rho_2 \leq \rho_3 \leq \rho_4, \tag{3.8.7}$$

and denote $z_1^{\mathrm{T}} = (x_1, \ldots, x_4)$, $z_2 = \sigma$.

With the subsystems (3.8.3) and (3.8.4) connect the functions

$$v_1(z_1) = c_1 z_1^{\mathrm{T}} z_1, \quad v_2(z_2) = c_2 z_2^2, \tag{3.8.8}$$

where $c_1, c_2 > 0$ are constants.

It is easy to verify that the following estimates hold:

$$\left. \frac{dv_1(z_1)}{dt} \right|_{(3.8.3)} \leq -2c_1 \rho_1 \|z_1\|^2, \tag{3.8.9}$$

$$\left. \frac{dv_2(z_2)}{dt} \right|_{(3.8.4)} \leq -2rpc_2 \|z_2\|^2, \tag{3.8.10}$$

$$\|\operatorname{grad} v_1(z_1)\| \leq 2c_1 \|z_1\|, \tag{3.8.11}$$

$$\|\operatorname{grad} v_2(z_2)\| \leq 2c_2 \|z_2\| \tag{3.8.12}$$

at all $z_1 \in R^4$, $z_2 \in R$. The norms of matrices (3.8.6), concordant with the Euclidean norm of vectors, are

$$\|C_{12}\| = 2, \quad \|C_{21}\| = \left(\sum_{i=1}^{4} \beta_i^2 \right)^{1/2}. \tag{3.8.13}$$

For further treatment we will use a corollary of Theorem 3.6.3.

Corollary 3.8.1 Let the equations of perturbed motion (3.8.3) and (3.8.4) be such that:

(1) for each subsystem (3.8.3) and (3.8.4) there exist functions $v_1(z_1)$ and $v_2(z_2)$ such that

$$c_{i1}\|z_i\|^2 \leq v_i(z_i) \leq c_{i2}\|z_i\|^2,$$

$$\left.\frac{dv_i(z_i)}{dt}\right|_{(\cdot)} \leq \sigma_i\|z_i\|^2, \quad \text{where} \quad (\cdot) = \begin{cases} (3.8.3), \\ (3.8.4); \end{cases}$$

(2) for the specified functions $v_i(z_i)$, $i = 1,2$, there exist positive constants c_{i3} for which

$$\|\mathrm{grad}\, v_i(z_i)\| \leq c_{i3}\|z_i\| \quad \text{at all} \quad z_i \in R^{n_i};$$

(3) at specified σ_i, $i = 1,2$, there exist a vector $\alpha = (\alpha_1, \alpha_2)^{\mathrm{T}}$ and a value $\mu^* \in M$ such that the matrix $S(\mu) = [s_{ij}(\mu)]$ with the elements

$$s_{ij}(\mu) = \begin{cases} \alpha_i \sigma_i & \text{at } i = j, \\ \frac{1}{2}\mu[\alpha_i c_{i3}\|C_{ij}\| + \alpha_j c_{j3}\|C_{ji}\|] & \text{at } i \neq j, \end{cases} \tag{3.8.14}$$

is negative definite at all $\mu < \mu^*$.

Then the equilibrium state (3.8.2) of the system (3.8.1) is globally μ-stable.

Taking into account (3.8.9)–(3.8.12), for the elements $s_{ij}(\mu)$ of the matrix $S(\mu)$ obtain the expressions according to (3.8.14):

$$s_{11} = -2\alpha_1 c_1 \rho_1, \quad s_{22} = -2\alpha_2 rpc_2,$$

$$s_{12} = s_{21} = 2\alpha_1 \mu c_1 + \alpha_2 \mu c_2 \left(\sum_{i=1}^{4} \beta_i^2\right)^{1/2}.$$

Now choose

$$a_1 = \frac{1}{4c_1}, \quad a_2 - \frac{1}{2c_2\left(\sum\limits_{i=1}^{4}\beta_i^2\right)^{1/2}}.$$

Here

$$s_{11} = -\frac{1}{2}\rho_1, \quad s_{22} = -\frac{rp}{\left(\sum\limits_{i=1}^{4}\beta_i^2\right)^{1/2}}, \quad s_{12} = s_{21} = \mu.$$

Therefore, the matrix $S(\mu)$ has the form

$$S(\mu) = \begin{pmatrix} -\dfrac{1}{2}\rho_1 & \mu \\ \mu & -\dfrac{rp}{\left(\sum\limits_{i=1}^{4}\beta_i^2\right)^{1/2}} \end{pmatrix}.$$

The matrix $S(\mu)$ is negative definite if at $\mu < \mu^*$ the following inequalities hold:

$$-\frac{1}{2}\rho_1 < 0, \quad \left(-\frac{1}{2}\rho_1\right)\left(-\frac{rp}{\left(\sum\limits_{i=1}^{4}\beta_i^2\right)^{1/2}}\right) - \mu^2 > 0.$$

Since $\rho_1 > 0$, the first inequality holds automatically. From the second inequality obtain

$$\mu^2\left(\sum\limits_{i=1}^{4}\beta_i^2\right)^{1/2} < \frac{1}{2}rp\rho_1 \tag{3.8.15}$$

at $\mu < \mu^*$.

The condition (3.8.15) is sufficient for the global exponential μ-stability of the motion of the system (3.8.1).

3.8.2 Indirect control of systems

Among the problems of automatic control, an important place is held by the problem of indirect control (see Lefshets [1] and others). Consider the equations of perturbed motion of such a system with small nonlinearities

$$\begin{aligned}
\frac{dx}{dt} &= Ax + \mu bf(\sigma), \\
\frac{d\sigma}{dt} &= -\rho\sigma - rf(\sigma) + \mu a^{\mathrm{T}}x,
\end{aligned} \tag{3.8.16}$$

where $x \in R^n$, $\sigma \in R$, A is a stable $(n \times n)$-matrix (Re $\lambda_j < 0$), $b \in R^n$, $\rho > 0$, $a \in R^n$, $\mu \in (0,1]$ and $f\colon R \to R$ and has the following properties:

(a) $f(\sigma) = 0$ if and only if $\sigma = 0$;

(b) $f(\sigma)$ is continuous on R;

(c) $0 < \sigma f(\sigma) \le k\sigma^2$ at all $\sigma \ne 0$, where $k = \mathrm{const} > 0$.

At $\mu = 0$ from the system (3.8.16) obtain the independent subsystems

$$\frac{dx}{dt} = Ax, \tag{3.8.17}$$

$$\frac{d\sigma}{dt} = -\rho\sigma - rf(\sigma). \tag{3.8.18}$$

The connection functions have the form

$$\begin{aligned}
g_1(x,\sigma,\mu) &= \mu bf(\sigma), \\
g_2(x,\mu) &= \mu a^{\mathrm{T}}x.
\end{aligned}$$

For the subsystems (3.8.17) and (3.8.18) construct the Lyapunov functions

$$v_1(x) = x^T B x, \tag{3.8.19}$$

where $A^T B + BA = -C$, C is some positive definite matrix, and

$$v_2(\sigma) = \frac{1}{2}\sigma^2. \tag{3.8.20}$$

For the function $v_i(\cdot)$, $i = 1, 2$, the following estimates hold:

$$c_{11}\|x\|^2 \leq v_1(x) \leq c_{12}\|x\|^2,$$

$$\frac{dv_1(x)}{dt}\Big|_{(3.8.17)} \leq -c_{13}\|x\|^2,$$

$$\|\operatorname{grad} v_1(x)\| \leq c_{14}\|x\| \quad \text{at all} \quad x \in R^n;$$

$$\frac{dv_2(\sigma)}{dt}\Big|_{(3.8.18)} \leq -\rho|\sigma|^2,$$

$$\|\operatorname{grad} v_2(\sigma)\| \leq |\sigma| \quad \text{at all} \quad \sigma \in R.$$

For the connection functions g_1 and g_2 the following estimates hold:

$$\|g_1(x, \sigma, \mu)\| \leq \mu k|b|\|x\|,$$

$$\|g_2(x, \mu)\| \leq \mu|a||\sigma|.$$

Choose the values

$$\alpha_1 = \frac{1}{k|b|} \quad \text{and} \quad \alpha_2 = \frac{c_{14}}{|a|} \tag{3.8.21}$$

and construct a matrix $S(\mu)$ with the elements

$$s_{ij}(\mu) = \begin{cases} \alpha_i \sigma_j & \text{at} \quad i = j, \\ \dfrac{1}{2}\left[\alpha_i c_{13}\|c_{ij}\| + \alpha_j c_{j3}\|c_{ji}\|\right] & \text{at} \quad i \neq j, \end{cases} \tag{3.8.22}$$

where $\|c_{ij}\| = \mu k|b|$, $\|c_{ji}\| = \mu|a|$.

Taking into account (3.8.21) and the estimates for the functions $v_1(x)$ and $v_2(\sigma)$, it is not difficult to find expressions for the matrix

$$S(\mu) = \begin{pmatrix} -\dfrac{c_{13}}{\mu k|b|} & c_{14} \\ c_{14} & -\dfrac{c_{14}\rho}{\mu|a|} \end{pmatrix}. \tag{3.8.23}$$

The matrix (3.8.23) is negative definite if

$$\mu^2 k < \frac{\rho c_{13}}{c_{14}|a||b|} \quad \text{at} \quad \mu < \mu^*, \quad \mu^* \in (0, 1]. \tag{3.8.24}$$

Due to the presence of a small parameter in the inequality (3.8.24), the range of values of the parameters of the system (3.8.16) may be extended.

Thus, since the function

$$v(x, \sigma) = \alpha_1 v_1(x) + \alpha_2 v_2(\sigma)$$

is positive definite and decrescent, under the inequality (3.8.24) the equilibrium state $(x^T, \sigma) = 0$ of the indirect control system (3.8.16) is globally asymptotically μ-stable.

3.8.3 Control system with an unstable free subsystem

Continue the study of the system (3.8.16). Using a nonsingular linear transformation (see Michel and Miller [1]) the system (3.8.16) can be reduced to the form

$$\frac{dx_1}{dt} = A_1 x_1 + \mu b_1 f(\sigma),$$

$$\frac{dx_2}{dt} = A_2 x_2 + \mu b_2 f(\sigma), \qquad (3.8.25)$$

$$\frac{d\sigma}{dt} = -\rho\sigma - rf(\sigma) + \mu a_1^T x_1 + \mu a_2^T x_2,$$

where $x_1 \in R^{n_1}$, $x_2 \in R^{n_2}$, A_1 is a constant $(n_1 \times n_1)$-matrix, A_2 is a constant $(n_2 \times n_2)$-matrix, $b_1 \in R^{n_1}$, $b_2 \in R^{n_2}$, $a_1 \in R^{n_1}$, $a_2 \in R^{n_2}$, $\mu \in (0, 1]$, $n_1 + n_2 = n$.

At $\mu = 0$ from the system (3.8.25) obtain three subsystems

$$\frac{dx_1}{dt} = A_1 x_1, \qquad (3.8.26)$$

$$\frac{dx_2}{dt} = A_2 x_2, \qquad (3.8.27)$$

$$\frac{d\sigma}{dt} = -\rho\sigma rf(\sigma). \qquad (3.8.28)$$

The connection functions have the form

$$g_1(x_1, x_2, \sigma, \mu) = \mu b_1 f(\sigma),$$
$$g_2(x_1, x_2, \sigma, \mu) = \mu b_2 f(\sigma),$$
$$g_3(x_1, x_2, \sigma, \mu) = \mu a_1^T x_1 + \mu a_2^T x_2.$$

Make the following assumptions about the subsystems (3.8.26) and (3.8.27):

(a) all eigenvalues of the matrix A_1 have positive real parts;

(b) the matrix A_2 is stable, that is, $\operatorname{Re}\lambda_j(A_2) < 0$, $j = 1, 2, \ldots, n_2$.

In addition, there exist functions $v_1 \colon R^{n_1} \to R$ and $v_2 \colon R^{n_2} \to R$, positive constants c_{ij}, $i=1,2$, $j = 1, 2, 3, 4$, such that

$$
\begin{aligned}
-c_{11}\|x_1\|^2 &\leq v_1(x_1) \leq -c_{12}\|x_2\|^2, \\
\left.\frac{dv_1(x_1)}{dt}\right|_{(3.8.26)} &\leq -c_{12}\|x_2\|^2, \\
\|\operatorname{grad} v_1(x_1)\| &\leq c_{14}\|x_1\|, \\
c_{21}\|x_2\|^2 &\leq v_2(x_2) \leq c_{22}\|x_2\|^2, \\
\left.\frac{dv_2(x_2)}{dt}\right|_{(3.8.27)} &\leq -c_{23}\|x_2\|^2, \\
\|\operatorname{grad} v_2(x_2)\| &\leq c_{24}\|x_2\|
\end{aligned}
\tag{3.8.29}
$$

at all $x_1 \in R^{n_1}$ and $x_2 \in R^{n_2}$.

For the subsystem (3.8.28) take the function $v_3(\sigma)$ in the form

$$
v_3(\sigma) = \frac{1}{2}\,\sigma^2.
$$

For this function obtain

$$
\begin{aligned}
\left.\frac{dv_3(\sigma)}{dt}\right|_{(3.8.28)} &\leq -\rho|\sigma|^2, \\
|\operatorname{grad} v_3(\sigma)| &\leq |\sigma|
\end{aligned}
\tag{3.8.30}
$$

at all $\sigma \in R$.

The constants a_{ij} from condition (3) of Assumption 3.6.5 for the system (3.8.25) have the form

$$
\begin{aligned}
a_{13} &= c_{14}\mu k\|b_1\|, \quad a_{23} = c_{24}\mu k\|b_2\|, \quad a_{31} = \mu\|a_1\|, \\
a_{32} &= \mu\|a_2\|, \quad a_{12} = a_{21} = a_{11} = a_{22} = a_{33} = 0.
\end{aligned}
\tag{3.8.31}
$$

Now apply Theorem 3.6.5 on μ-instability. Taking into account the conditions (3.8.29)–(3.8.31), for the elements of the matrix $S(\mu)$ obtain the expression

$$
S(\mu) = \begin{pmatrix} c_{13} & 0 & -c_{14}\mu k\|b_1\| \\ 0 & c_{23} & -c_{24}\mu k\|b_2\| \\ -\mu\|a_1\| & -\mu\|a_2\| & \rho \end{pmatrix}.
\tag{3.8.32}
$$

The matrix (3.8.32) is positive definite at $\mu < \mu^*$, if and only if

$$
\mu^2 k < \frac{c_{13}c_{23}\rho}{c_{23}c_{14}\|a_1\|\|b_1\| + c_{13}c_{24}\|a_2\|\|b_2\|}, \quad \mu^* \in (0, 1].
\tag{3.8.33}
$$

Thus, Theorem 3.6.5 implies that the equilibrium state $(x_1^{\mathrm{T}}, x_2^{\mathrm{T}}, \sigma) = 0$ of the system (3.8.25) is μ-unstable at all admissible nonlinearities of f, if the inequality (3.8.33) holds.

3.9   Comments and References

The dynamic analysis of nonlinear weakly connected systems is conducted in this chapter on the basis of one variant of the comparison technique that was substantiated in Chapter 1. As is known, the advantage of such approach lies in the fact that it allows us to make a conclusion about the stability (instability) of solutions of the initial system via the analysis of the properties of solutions of a scalar equation and the properties of auxiliary functions used for the transformation of the initial equations.

3.2. The statement of the problem on the stability of nonlinear weakly connected systems in terms of two different measures for four types of connection functions between the subsystems is discussed. The similarity and the difference between the statement and the problem of stability under continuous perturbations are described (see Gorshin [1], Duboshin [1, 2], Malkin [1], Martynyuk [12, 13]).

3.3. Theorems 3.3.1 – 3.3.5 are new. To obtain them, some of the results of the monographs of Lyapunov [1], Malkin [2], and Lakshmikantham, Leela, and Martynyuk [1] were used.

3.4. Theorem 3.4.1 is new. It was obtained through the application of the perturbed Lyapunov function and the scalar variant of the comparison principle. Note that starting from the work of Corduneanu [2] the principle of comparison with the scalar and vector Lyapunov functions has been applied in many lines of investigation (see Matrosov [2], Rouche, Habets, and Laloy [1], Rao [1], and others).

3.5. Theorems 3.5.1 and 3.5.2 are new. The idea of the dynamic analysis of an individual subsystem in a complex system was proposed in the monograph of Martynyuk [6]. In this section, the perturbed Lyapunov function and the technique of comparison with the perturbed Lyapunov function are applied (cf. Lakshmikantham and Leela [1]).

3.6. The results have been obtained on the basis of specific assumptions on the dynamic properties of independent subsystems and the functions of connection between them. The sufficient conditions for the respective types of stability are formulated in terms of sign definiteness of special matrices. All the results are new for the system (3.2.1). They were obtained by using some results of the monograph of Michel and Miller [1]. The results of the analysis of stability of large-scale systems that do not contain a small parameter can be found in many journals.

3.7. The polystability of motion presents a new line of investigation in the nonlinear dynamics of systems (see Aminov and Sirazetdinov [1], Martynyuk [11, 15]). This section is based on the article of V.A. Martynyuk [1]. This problem is related to the problem of stability with respect to a part of variables (see Rumiantsev [1], Vorotnikov [1], Corduneanu [3], Martynyuk [14], Peiffer and Rouche [1], Hatvani [1]).

3.8. Some applications of the general results are described. Here the systems of automatic control are studied whose motion equations are given in the monographs of Aizerman and Gantmacher [1], Lefschetz [1], and Lurie [1] (see Chapter 2). The sufficient conditions for the μ-stability were obtained in algebraic form (cf. Michel and Miller [1]). The results of this section have not been published before.

Chapter 4

Stability of Weakly Perturbed Systems

4.1 Introductory Remarks

As far as in 19th century Delaunay and Hill noticed that averaging along the solution of a degenerate system in celestial mechanics problems provided results that corresponded to real phenomena. The contribution to Delaunay's theory made by Tissarand formed the conceptual background of the new methods of celestial mechanics developed by Poincaré.

The investigation of the stability of systems with a small parameter which was started in Chapter 3 is continued here. In contrast to Chapter 3, here the ideas of the method of averaging of the nonlinear mechanics are added to the direct Lyapunov method. In some instances such an approach makes it possible to study classes of systems with a small parameter under new wider assumptions on the properties of solutions of generating equations.

In Section 4.2, the stable-like properties of solutions of a weakly perturbed nonlinear system are investigated. Analogues of the main theorems of the direct Lyapunov method for the given class of systems are cited.

In Section 4.3, the investigation of the same class of systems is continued, but on a finite time interval. The estimate of the interval on which the solution does not leave the ε-neighborhood of the stationary point is shown.

Section 4.4 contains the theorems on the stability of a weakly perturbed system by Lyapunov and the stability on a finite time interval when a special mean is calculated along solutions of a limiting system corresponding to the generating system.

Section 4.5 contains the results of the analysis of stability of weakly connected large-scale systems with nonasymptotically stable subsystems.

Section 4.6 contains the generalization of one of the theorems of Section 4.2 for the case of stability with respect to a part of variables.

Section 4.7 contains some applications of the general results to the analysis of oscillatory systems.

Section 4.8 contains comments and a bibliography.

4.2   Averaging and Stability

4.2.1   Problem and auxiliary results

Consider the system of differential equations

$$\frac{dx}{dt} = f(t,x) + \mu g(t,x), \quad x(t_0) = x_0, \tag{4.2.1}$$

where $x \in R^n$, the vector functions $f(t,x)$ and $g(t,x)$ are defined and continuous in the domain $\Omega = \{x \in R^n, \ t \in J: \|x\| < H, \ H = \text{const} > 0\}$. It is assumed that at $(t_0, x_0) \in \text{int} \, \Omega$ a solution of the system exists and is unique on the interval $J = [t_0, \infty)$. In addition, the vector function $f(t,x)$ in the domain Ω satisfies the Lipschitz condition with respect to x

$$\|f(t,x') - f(t,x'')\| \leq L\|x' - x''\|, \quad L = \text{const} > 0. \tag{4.2.2}$$

It is assumed that the degenerate system

$$\frac{dx}{dt} = f(t,x), \quad x(t_0) = x_0 \tag{4.2.3}$$

has the state of equilibrium $x = 0$ $(f(t,0) = 0)$ which is stable uniformly with respect to t_0 and for it the general solution $\overline{x}(t) = \overline{x}(t, t_0, x_0)$, $\overline{x}(t_0) = x_0$, $(t_0, x_0) \in \text{int} \, \Omega$ is known.

For the study of the system (4.2.1) we will use the mean of the scalar product of the gradient of the Lyapunov function $v_0(t,x)$ of the degenerate system and the perturbation vector $g(t,x)$, calculated along solutions of the system (4.2.3),

$$\Theta_0(t_0, x_0) = \lim_{T \to \infty} \frac{1}{T} \int_{t_0}^{t_0+T} \varphi(t, \overline{x}(t)) \, dt, \tag{4.2.4}$$

where $\varphi(t,x) = \left(\partial v_0/\partial x\right)^{\mathrm{T}} g(t,x)$ exists in the domain Ω.

Now we will need the following auxiliary statements.

Lemma 4.2.1 *Let the function $u(t)$ be continuous and nonnegative on the interval $[\alpha, \beta]$ and satisfy the inequality*

$$u(t) \leq \int_{\alpha}^{t} a(\tau)u(\tau) \, d\tau + f(t),$$

where $a(\tau)$ is a function nonnegative and integrable on $[\alpha, \beta]$ and $f(t)$ is a function bounded on $[\alpha, \beta]$.

Then the following inequality holds:

$$u(t) \leq \sup_{\alpha \leq t \leq \beta} |f(t)| \exp\left(\int\limits_{\alpha}^{t} a(\tau) \, d\tau \right).$$

This estimate is a minor generalization of the Bellman–Gronwall theorem.

Lemma 4.2.2 *Let the vector function $f(t,x)$ satisfy the Lipschitz condition with respect to x and let there exist a summable function $M(t)$ and a constant M_0 such that in the domain Ω on any finite interval $[t_1, t_2]$ the following inequalities hold:*

$$\|g(t,x)\| \leq M(t), \quad \int\limits_{t_1}^{t_2} M(t) \, dt \leq M_0(t_2 - t_1). \tag{4.2.5}$$

Then for the norm of difference of the solutions $x(t, t_0, x_0)$ of the system (4.2.1) and $\overline{x}(t, t_0, x_0)$ of the system (4.2.3) at all $t \in [t_0, t_0 + l]$ the following inequality holds:

$$\|x(t) - \overline{x}(t)\| \leq \mu M_0 l \exp(Ll). \tag{4.2.6}$$

The estimate (4.2.6) is obtained through direct application of the Bellman lemma to the integral equation obtained from the equation (4.2.1).

Lemma 4.2.3 *Assume that the vector functions $f(t,x)$ and $g(t,x)$ satisfy the Lipschitz condition with respect to x with a constant L, and in addition, $g(t,0) = 0$. Then for the solutions $x(t, t_0, x_0)$ and $\overline{x}(t, t_0, x_0)$ of the systems (4.2.1) and (4.2.3) the following inequalities hold:*

(a) $\|x(t,\mu)\| \leq \|x_0\| \exp[L(1+\mu)(t - t_0)] \leq \|x_0\| Q_x(t - t_0),$

(b) $\|\overline{x}(t)\| \leq \|x_0\| \exp[L(t - t_0)] = \|x_0\| Q_{\overline{x}}(t - t_0),$

(c) $\|x(t,\mu) - \overline{x}(t)\| \leq \mu \|x_0\| \{\exp[L(t - t_0)] - 1\}$
$\times \exp[L(\mu + 1)(t - t_0)] \leq \mu \|x_0\| Q(t - t_0).$

$$\tag{4.2.7}$$

Here the obvious notation is introduced, and in the functions $Q_x(t - t_0)$ and $Q(t - t_0)$ it is assumed that $\mu = \mu^*$.

Proof From the equations (4.2.1) and (4.2.3), using the conditions of Lemma 4.2.3, obtain

$$\|x(t, t_0, x_0)\| \leq \|x_0\| + \int\limits_{t_0}^{t} L(\mu + 1) \|x(t, t_0, x_0)\| \, dt,$$

$$\|\overline{x}(t, t_0, x_0)\| \leq \|x_0\| + \int\limits_{t_0}^{t} L \|\overline{x}(t, t_0, x_0)\| \, dt.$$

Hence, using Lemma 4.2.1, obtain inequalities (a) and (b) from (4.2.7).
Inequality (c) from (4.2.7) follows from

$$\|x(t,\mu) - \overline{x}(t)\| \leq \int_{t_0}^{t} [\|f(t,x(t)) - f(t,\overline{x}(t))\| + \mu\|g(t,x(t))$$

$$- g(t,\overline{x}(t))\|] \, dt + \mu \int_{t_0}^{t} \|g(t,\overline{x}(t))\| \, dt$$

$$\leq \int_{t_0}^{t} L(1+\mu)\|x(t) - \overline{x}(t)\| dt + \mu \int_{t_0}^{t} L\|\overline{x}(t)\| dt$$

$$\leq \int_{t_0}^{t} L(1+\mu)\|x(t) - \overline{x}(t)\| dt + \mu\|x_0\|\{\exp[L(t-t_0)] - 1\}.$$

The lemma is proved.

Recall some notation. The distance from a point x to a set M will be denoted as follows:

$$\rho(x, M) = \inf(\|x - x'\|, \ x' \in M).$$

We will use continuously differentiable Lyapunov functions $v(t,x)$ defined in a domain Ω. Let $v^*(x)$ denote a nonpositive function defined and continuous in a domain D. The set of points $x \in D$ for which $v^*(x) = 0$ will be denoted by $E(v^* = 0)$. Similarly, introduce the notation of a nonnegative function $w^*(x)$ and a set $E(w^* = 0)$.

4.2.2   Conditions for stability

The conditions for the stability of solutions of the system (4.2.1) were obtained in the case of "neutrality" of the shortened system (4.2.3) on the basis of the mean (4.2.4).

The main requirement ensuring the stability of a trivial solution is the negativeness of the mean outside an arbitrarily small neighborhood of the point $x = 0$.

Here we will obtain the conditions for the μ-stability under weaker limitations on the mean (4.2.4).

Definition 4.2.1   The mean $\Theta_0(t_0, x_0)$ is less than zero in the set $E(v^* = 0)$, if for any numbers η and ε, $0 < \eta < \varepsilon < H$, there exist positive numbers $r(\eta, \varepsilon)$ and $\delta(\eta, \varepsilon)$ such that $\Theta_0(t_0, x_0) < -\delta(\eta, \varepsilon)$ at $\eta \leq \|x_0\| \leq \varepsilon$, $\rho(x_0, E(v^* = 0)) < r(\eta, \varepsilon)$ for all $t_0 \in J$.

Theorem 4.2.1 *Let the following conditions be satisfied:*

(1) *in the domain Ω there exists a positive definite decrescent function $v(t,x)$ such that*

$$\frac{\partial v}{\partial t} + \left(\frac{\partial v}{\partial x}\right)^{\mathrm{T}} f(t,x) \leq v^*(x) \leq 0;$$

(2) *there exist summable functions $M(t)$, $F(t)$, $N(t)$, constants M_0, F_0, and N_0, and a function $\chi(\beta) \in R$ such that the following inequalities hold:*

$$\varphi(t,x) \leq N(t), \quad \int_{t_1}^{t_2} N(t)\,dt \leq N_0(t_2 - t_1) \qquad (4.2.8)$$

at $x \in D \setminus E(v^* = 0)$, $t \in J$ and

$$\|g(t,x)\| \leq M(t), \quad \int_{t_1}^{t_2} M(t)\,dt \leq M_0(t_2 - t_1),$$

$$\|\varphi(t,x') - \varphi(t,x'')\| \leq \chi(\|x' - x''\|)F(t), \qquad (4.2.9)$$

$$\int_{t_1}^{t_2} F(t)\,dt \leq F_0(t_2 - t_1)$$

at $(t,x) \in \Omega$ on any finite interval $[t_1, t_2]$;

(3) *uniformly with respect to $(t_0, x_0) \in \Omega$ there exists a mean $\Theta_0(t_0, x_0)$;*

(4) *the mean $\Theta_0(t_0, x_0)$ is less than zero in the set $E(v^* = 0)$.*

Then the solution $x = 0$ of the system (4.2.3) is μ-stable.

Proof Let $\varepsilon \in (0, H)$ and $t_0 \in R_+$ be specified. We will show the method used for choosing $\eta(\varepsilon)$ and $\mu_0(\varepsilon)$, which do not depend on t_0. Assume that the conditions of Theorem 4.2.1 are satisfied. For the positive definite decrescent function $v(t,x)$ there exist functions $a(r)$ and $b(r)$ from the K class, such that in the domain Ω the following condition is satisfied:

$$a(\|x\|) \leq v(t,x) \leq b(\|x\|). \qquad (4.2.10)$$

In view of (4.2.10) all points of the moving surface $v(t,x) = a(\varepsilon/2)$ will satisfy the condition

$$b^{-1}(a(\varepsilon/2)) \leq \|x\| \leq \varepsilon/2 \qquad (4.2.11)$$

for all $t \in R_+$. Assume that $\eta(\varepsilon) = b^{-1}(a(\varepsilon/2))$.

Consider the solution $x(t, \mu)$ of the system (4.2.1) with the initial conditions $\|x_0\| < \eta(\varepsilon)$. Assume that it left the domain $\|x\| < \eta(\varepsilon)$ and at a point of time $t = t_0$ crossed the surface $v(t,x) = a(\varepsilon/2)$ in a point x'_0. For this point the inequality (4.2.11) holds and therefore, in view of condition (4) of the theorem for $\varepsilon/2$ and $\eta(\varepsilon)$, there exist numbers $r(\eta, \varepsilon/2) > 0$ and $\delta(\eta, \varepsilon/2) > 0$ such that one of the following conditions is satisfied:

$$\rho(x'_0, E(v^* = 0)) \geq r(\eta, \varepsilon/2), \quad \overline{\varphi}(t'_0, x'_0) < -\delta(\eta, \varepsilon/2).$$

Consider some properties of the solution $x(t, \mu)$:

(a) Let the following conditions be satisfied at a point of time τ:

$$v(\tau, x(\tau)) = a(\varepsilon/2), \quad \rho(x(\tau), E(v^* = 0)) \geq r(\eta, \varepsilon/2).$$

Study the behavior of the function $v(t, x)$ along the solution $x(t, \mu)$ of the system (4.2.1). Integrating the expression of the full derivative of the function $v(t, x)$, taking into account the system (4.2.1), for $t > \tau$ obtain

$$v(t, x(t)) \leq v(\tau, x(\tau)) + \int_\tau^t v^*(x(t))\, dt + \mu \int_\tau^t \varphi(t, x(t))\, dt. \qquad (4.2.12)$$

In this situation there exists a positive number $\gamma = \inf_{x \in P} |v^*(x)|$, where

$$P = \{x: \ \rho(x, E(v^* = 0)) \geq r/2, \quad \eta \leq \|x\| \leq \varepsilon/2\}.$$

Choosing $\mu < \mu'_0 = \gamma/2N_0$ and using the inequalities (4.2.8) for all $t > \tau$, for which the conditions $\|x(\tau)\| \geq \eta$ and $\rho(x(t), E(v^* = 0)) \geq r/2$ are satisfied, from the inequality (4.2.12) obtain

$$v(t, x(t)) \leq a(\varepsilon/2) - \frac{\gamma}{2}(t - \tau). \qquad (4.2.13)$$

It implies that at $\mu < \mu'_0$ the function $v(t, x(t))$ is not increscent, which means that the solution $x(t, \tau, x(\tau))$ in view of the inequality (4.2.11) will not leave the domain $\|x\| < \varepsilon/2$, at least until the inequality $\rho(x(t), E(v^* = 0)) \geq r/2$ is violated.

(b) Let the following conditions be satisfied at a point τ:

$$v(\tau, x(\tau)) = a(\varepsilon/2), \quad \rho(x(\tau), E(v^* = 0)) < r(\eta, \varepsilon/2).$$

In this event, as a result of condition (4) of Theorem 4.2.1, the following inequality holds:

$$\Theta_0(\tau, x(\tau)) < \delta(\eta, \varepsilon/2). \qquad (4.2.14)$$

Therefore, we will estimate the change of the function $v(t, x)$ along the solution $x(t, \mu)$. The first integral in the inequality (4.2.12) will be neglected, and the second one will be represented in the form

$$\int_\tau^t \varphi(t, x(t))\, dt = \int_\tau^t [\varphi(t, x(t)) - \varphi(t, \overline{x}(t))]\, dt + \int_\tau^t \varphi(t, \overline{x}(t))\, dt. \qquad (4.2.15)$$

Here $\bar{x}(t)$ is a solution of the system (4.2.3).

Condition (3) of the theorem implies that there exists a function $\varkappa(t)$ such that $\lim\limits_{t\to\infty} \varkappa(t) = 0$, and the last integral in (4.2.15) can be represented in the form

$$\int_{\tau}^{t} \varphi(t, \bar{x}(t))\, dt = (t - \tau)\, [\bar{\varphi}(\tau, x(\tau)) + \varkappa(t)]. \qquad (4.2.16)$$

Choose the time interval l so large that for $t > \tau + l$ the condition

$$|\varkappa(t)| < \delta(\eta, \varepsilon/2)/4 \qquad (4.2.17)$$

will be satisfied, and construct estimates on the interval $[\tau, \tau + 2l]$. Choosing

$$\mu_1 = \varkappa^{-1}(\delta/4F_0)/2M_0 l \exp(2Ll)$$

and using Lemma 4.2.2 for $\mu < \mu_1$ and $t \in [\tau, \tau + 2l]$, obtain

$$\|x(t) - \bar{x}(t)\| < \chi^{-1}(\delta/4F_0). \qquad (4.2.18)$$

From the inequality (4.2.9) and the estimate (4.2.18) it follows that

$$\int_{\tau}^{t} |\varphi(t, x(t)) - \varphi(t, \bar{x}(t))|\, dt < \frac{\delta}{4}(t - \tau) \qquad (4.2.19)$$

for all $t \in [\tau, \tau + 2l]$ and $\mu < \mu_1$. Choose μ_2 so that at $\mu < \mu_2$ on the interval $[\tau, \tau + 2l]$ the solution $x(t, \mu)$ would not leave the domain $\|x\| < \varepsilon$. For this purpose, taking into account that $\|\bar{x}(t, \mu)\| \leq \varepsilon/2$, it is sufficient to demand that $\|x(t, \mu) - \bar{x}(t, \mu)\| < \varepsilon/2$. Then from the inequality (4.2.6) obtain

$$\mu_2 = \frac{\varepsilon}{4M_0 l \exp(2lL)}.$$

Choose $\mu_0'' = \min(\mu_1, \mu_2)$, then at $\mu < \mu_0''$ for $t \in [\tau, \tau + 2l]$ the condition $\|x(t, \mu)\| < \varepsilon$ and the estimate (4.2.19) will hold. Substituting (4.2.16) and (4.2.19) into (4.2.11) and using the inequalities (4.2.14) and (4.2.17), for $t \in [\tau + l, \tau + 2l]$ and $\mu < \mu_0''$ obtain

$$\int_{\tau}^{t} \varphi(t, x(t))\, dt \leq -\frac{\delta}{4}(t - \tau). \qquad (4.2.20)$$

As is obvious from (4.2.20), the last integral in the expression (4.2.12), at least starting from the point $t = \tau + l$, becomes negative, which means that the solution $x(t, \mu)$, having left $v(t, x) = a(\varepsilon/2)$, due to the chosen $\mu < \mu_0''$, will remain at $t \in [\tau, \tau + 2l]$ in the domain $\|x\| < \varepsilon$ and at some point of time $t^* \in [\tau, \tau + 2l]$ return to the domain bounded by the surface $v(t, x) = a(\varepsilon/2)$.

Choose $\mu_0 = \min(\mu_0', \mu_0'')$. At a point of time t_0' for the solution $x(t, \mu)$ of the system (4.2.1) one of the above conditions (a) or (b) is satisfied. Choosing a time interval depending on what case holds, we can show that the solution of the system (4.2.1) in the finite point of the interval lies within the domain bounded by the surface $v(t, x) = a(\varepsilon/2)$. Continuing this solution until the intersection with the above-mentioned surface, we obtain the initial point of the next interval, in which case (a) or case (b) will hold again. Since in both cases the estimates are uniform with respect to τ and $x(\tau)$, on all the following intervals either (4.2.13) or (4.2.20) will hold, which means that $x(t, \mu)$ for all $t \geq t_0$ will remain in the domain $\|x\| < \varepsilon$. The numbers η, μ_0 were chosen without regard to t_0.

Theorem 4.2.1 is proved.

Example 4.2.1 Study the equilibrium state $x = 0$ of the system of equations

$$\frac{dx_1}{dt} = -x_1 + x_2 + \mu(x_1^3 - ax_2^3),$$

$$\frac{dx_2}{dt} = \mu[a(x_1 + x_2)\cos t + (x_1^3 - ax_2^3)], \quad a = \text{const} > 0. \tag{4.2.21}$$

The derivative of the function $v(x) = x_2^2 + (x_1 - x_2)^2$ in view of the shortened system

$$\frac{dx_1}{dt} = -x_1 + x_2, \qquad \frac{dx_2}{dt} = 0 \tag{4.2.22}$$

has the form $\dfrac{dv}{dt}(x) = -2(x_1 - x_2)^2 = v^*(x) \leq 0$.

Having calculated the mean (4.2.4) along the solutions $\bar{x}_2 = x_{20}$, $\bar{x}_1 = x_{20} + (x_{10} - x_{20})\exp[-(t - t_0)]$ of the system (4.2.22), obtain

$$\Theta_0(x_0) = 2x_{20}^4(1 - a). \tag{4.2.23}$$

It is clear that the mean $\Theta_0(x_0)$ is less than zero in the set $E(v^* = 0) = \{x : x_1 = x_2\}$ at $a > 1$.

Thus, at $a > 1$ the solution $x = 0$ of the system (4.2.21) is μ-stable. Note that the mean (4.2.23) vanishes over the set $x_{20} = 0$ in the neighborhood of the point $x = 0$, and the derivative of the function $v(x) = x_2^2 + (x_1 - x_2)^2$ in view of the system (4.2.21) is an alternating function in this case.

4.2.3 Conditions of instability

The instability of solutions of the system (4.2.1) can be studied on the basis of Gorshin's theorem [1] on the instability under continuous perturbations, in which the instability of the equilibrium state of a shortened system is required. In the monograph by Khapaev [3] the cases were investigated when a shortened system is "neutral". An essential condition was the condition for the positiveness of the mean (4.2.3) in the domain $v > 0$.

In this subsection the conditions for the instability of the Chetaev theorem type were obtained under wider assumptions on the properties of solutions of the generating system.

Recall that a domain $v > 0$ is a neighborhood of the point $x = 0$ ($x \in D$), where the function $v(t, x)$ takes on positive values. It is assumed that this domain is bounded by the surface $v = 0$ and exists for all $t \in R_+$ at arbitrarily small x.

The mean $\Theta_0(t_0, x_0)$ is above zero in the sets $E_t(\Theta_0 = 0)$ of the domain $v > 0$, if for any positive λ, however small it may be, there exist positive numbers $\delta(\lambda)$ and $\chi(\lambda)$ such that at $t_0 \in R_+$ and $x_0 \in D$ satisfying the conditions $v(t_0, x_0) > \lambda$, $\dot{v}(t_0, x_0) \leq \chi(\lambda)$ the inequality $\Theta_0(t_0, x_0) > \delta(\lambda)$ holds.

Let $\{v_\tau > 0\}$ denote the intersection between the domain $v > 0$ and the plane $t = \tau$.

Theorem 4.2.2 *Let there exist a function $v(t, x)$ bounded in the domain $v > 0$, and, in addition,*

(1) *in the domain $v > 0$ the inequality $\dot{v}(t, x) \geq 0$ holds;*

(2) *there exist summable functions $M(t)$, $N(t)$, $F(t)$, constants M_0, N_0, F_0, and a function $\chi(\beta) \in K$, such that*

$$\|g(t, x)\| \leq M(t), \quad \int_{t_1}^{t_2} M(t)\, dt \leq M_0(t_2 - t_1),$$

$$\|\varphi(t, x') - \varphi(t, x'')\| \leq F(t)\chi(\|x' - x''\|), \qquad (4.2.24)$$

$$\int_{t_1}^{t_2} F(t)\, dt \leq F_0(t_2 - t_1)$$

in the domain $v > 0$ and

$$\varphi(t, x) \geq -N(t), \quad \int_{t_1}^{t_2} N(t)\, dt \leq N_0(t_2 - t_1) \qquad (4.2.25)$$

at $x \in \{v > 0\} \setminus (E_t(\dot{v} = 0) \cap \{v > 0\})$, $t \in J$ on any finite interval $[t_1, t_2]$;

(3) *uniformly with respect to t_0, x_0 in the domain $v > 0$ there exists a mean $\Theta_0(t_0, x_0)$;*

(4) *the mean $\Theta_0(t_0, x_0)$ is above zero in the sets $E_t(\dot{v} = 0)$ in the domain $v > 0$.*

Then the solution $x = 0$ of the system (4.2.1) is μ-unstable.

Proof Specify $\varepsilon \in (0, H)$ and $t_0 \in R_+$. Prove that however small x_0 may be chosen, one can always find an arbitrarily small number $\mu_0 > 0$ such that for the solution $x(t, \mu)$ of the system (4.2.1) at $\mu < \mu_0$ at some point of time $t^* > t_0$ the condition $\|x(t^*, \mu)\| = \varepsilon$ will be satisfied.

For a function $v(t, x)$ bounded in the domain $v > 0$ one can find a constant ω such that at $\|x\| < \varepsilon$ the following inequality would hold:

$$v(t, x) < \omega. \tag{4.2.26}$$

We will choose the value x_0 as small as we please, provided that the point x_0 belongs to the domain $v > 0$. Then there will exist such a positive number α, that the following condition will be satisfied:

$$v(t_0, x_0) > \alpha. \tag{4.2.27}$$

Under hypoghesis (4) of Theorem 4.2.2, for $\alpha > 0$ there exist positive numbers $\delta(\alpha)$ and $\chi(\alpha)$ such that one of the following inequalities holds:

$$\Theta_0(t_0, x_0) > \delta(\alpha), \quad \dot{v}(t_0, x_0) > \xi(\alpha).$$

Consider some properties of the solution $x(t, \mu)$ of the system (4.2.1).

(a) Let the following conditions be satisfied at a point τ:

$$v(\tau, x(\tau)) > \alpha, \quad \dot{v}(\tau, x(\tau)) > \xi(\alpha).$$

Consider the behavior of the function $v(t, x)$ along the solution $x(t, \mu)$. Integrating the expression of full derivative of the function $v(t, x)$, along solutions of the system (4.2.1) obtain

$$v(t, x(t)) = v(\tau, x(\tau)) + \int_\tau^t v(t, x(t))\, dt + \mu \int_\tau^t \varphi(t, x(t))\, dt. \tag{4.2.28}$$

Using the inequalities (4.2.25), at $\mu < \mu_0' = \xi(\alpha)/2N_0$ from the expression (4.2.28) find the estimate

$$v(t, x(t)) > \alpha + \frac{\xi}{2}(t - \tau) \tag{4.2.29}$$

for those t at which for the solution $x(t, \mu)$ the condition $\dot{v}(t, x(t)) > \xi(\alpha)$ is satisfied, that is, the inequality $v(t, x(t)) > \alpha$ for those points of time will not be violated. For the solution $x(t, \mu)$ the following conditions cannot be satisfied:

$$v(t, x(t)) > \xi(\alpha), \quad \|x(t, \mu)\| < \varepsilon \tag{4.2.30}$$

within the time interval $T = 2(\Omega - \alpha)/\xi(\alpha)$. Assume that those conditions are satisfied. Then from the inequality (4.2.29) for the point of time $\tau + T$ we obtain $v(\tau + T, x(\tau + T)) > \Omega$, which contradicts the condition for the boundedness of the function $v(t,x)$. The contradiction implies that there exists a point of time from the interval $(\tau, \tau + T)$, at which one of the inequalities (4.2.30) is violated. The violation of the second inequality from (4.2.30) means the instability. Assume that the first inequality of (4.2.30) is violated.

(b) Let the conditions $v(\tau, x(\tau)) > \alpha$, $\dot{v}(\tau, x(\tau)) \leq \chi(\alpha)$ be satisfied at a point of time τ. In this case, in view of condition (4) of Theorem 4.2.2 the following inequality will hold:

$$\Theta_0(\tau, x(\tau)) > \delta(\alpha). \tag{4.2.31}$$

Now we will estimate the change of the function $v(t,x)$ along the solution $x(t, \mu)$, using the mean $\Theta_0(\tau, x(\tau))$.

In the expression (4.2.28) neglect the first integral and represent the second one in the form

$$\int_\tau^t \varphi(t, x(t))\, dt = \int_\tau^t [\varphi(t, x(t)) - \varphi(t, \bar{x}(t))]\, dt + \int_\tau^t \varphi(t, \bar{x}(t))\, dt, \tag{4.2.32}$$

where $\bar{x}(t) = \bar{x}(t, \tau, x(\tau))$ is a solution of the system (4.2.3).

Represent the last integral from (4.2.32) in the form

$$\int_\tau^l \varphi(t, \bar{x}(t))\, dt = (t - \tau)[\Theta_0(\tau, x(\tau)) + \varkappa(t)]. \tag{4.2.33}$$

In view of condition (3) of the theorem, the function $\varkappa(t)$ is such that $\lim\limits_{t \to \infty} \varkappa(t) = 0$. Estimate the summands in the relation (4.2.32) on the interval $[\tau, \tau + 2l + T]$, where l will be chosen so large that for $t > \tau + l$ the following condition would be satisfied:

$$|\varkappa(t)| < \delta(\alpha)/4. \tag{4.2.34}$$

According to Lemma 4.2.2 for $t \in [\tau, \tau + 2l + T]$ find

$$\|x(t) - \bar{x}(t)\| \leq \mu M_0(2l + T) \exp[(2l + T)L]. \tag{4.2.35}$$

Choose

$$\mu_1 = \frac{\chi^{-1}(\delta(\alpha)/4F_0)}{M_0(2l + T) \exp[(2l + T)L]}.$$

Using the inequality (4.2.35) for $\mu < \mu_1$ and $t \in [\tau, \tau + (2l + T)]$, obtain

$$\int_\tau^t [\varphi(t, x(t)) - \varphi(t, \bar{x}(t))]\, dt < \frac{\delta(\alpha)}{4}(t - \tau). \tag{4.2.36}$$

Using (4.2.33), (4.2.34), and (4.2.36), for the integral (4.2.32) at a point of time $t \in [\tau + 2l, \tau + 2l + T]$ obtain the inequality

$$\int_\tau^t \varphi(t, x(t)) \, dt > \delta l.$$

Since all the requirements of Theorem 4.2.2 are satisfied in the domain $v > 0$, it is necessary to ensure that the solution $x(t, \tau, x(\tau))$ on the interval $[\tau, \tau + 2l + T]$ will not cross the surface $v = 0$ and will not leave the domain $v > 0$. For this purpose, impose another limitation on μ. Choose μ_2 so that at $t \in [\tau, \tau + 2l + T]$ the following condition would be satisfied:

$$\mu_2 \int_\tau^t \varphi(t, x(t)) dt \le \frac{\alpha}{2}.$$

Using the expression (4.2.32) and the inequality (4.2.36), this condition can be written as follows:

$$\mu_2 \left[\int_\tau^t \varphi(t, \overline{x}(t)) \, dt + \frac{\delta(\alpha)}{4}(t - \tau) \right] \le \frac{\alpha}{2}$$

at all $t \in [\tau, \tau + 2l + T]$, $\tau \in J$. Then at $\mu < \mu_0' = \min(\mu_1, \mu_2)$ on the interval $[\tau, \tau + 2l + T]$ the solution $x(t, \tau, x(\tau))$ will not leave the domain $v > 0$, and at the point $t \in [\tau + 2l, \tau + 2l + T]$ the following condition will be satisfied:

$$v(t, x(t)) > \alpha + \mu\delta l > \alpha. \qquad (4.2.37)$$

Choose $\mu_0 = \min(\mu_0', \mu_0'')$ and consider the sequence of points of time $t_i = t_0 + i(2l + T)$, $i = 1, 2, \ldots$. At the initial point of time t_0 condition (a) or (b) will be satisfied. If condition (1) is satisfied, then according to the proved property at $\mu < \mu_0$ on the solution of the system (4.2.1) $x(t, t_0, x_0)$ at some point of time t_0' from the interval $(t_0, t_0 + T)$ the conditions $v(t_0', x(t_0')) > \alpha$, $\Theta_0(t_0', x(t_0')) > \delta(\alpha)$ will be satisfied. Then, according to property (b), for the point of time $t_1 = t_0 + 2l + T$ the estimate (4.2.37) will hold. If at the initial point of time condition (b) is satisfied, then at the point t_1 we will also obtain the estimate (4.2.37). On all further intervals $[t_i, t_{i+1}]$ we will obtain similar cases, that is, on each interval the function $v(t, x(t))$ increases at least by $\mu\delta l > 0$ ($\mu \in (0, \mu_0)$). Let k be the smallest integer satisfying the condition $k \ge (\omega - \alpha)/\mu\delta l$.

Assume that at $t \in [t_0, t_k]$, where $t_k = t_0 + k(2l + T)$, the solution $x(t, \mu)$ lies in the domain $\|x\| < \varepsilon$. Taking into account the above, for a point of time t_k obtain $v(t_k, x(t_k)) > \Omega$, which contradicts the condition for the boundedness of the function $v(t, x)$. The obtained contradiction implies that there exists a point of time t^* from the interval (t_0, t_k), at which the condition $\|x(t^*, t_0, x_0)\| = \varepsilon$ is satisfied.

The theorem is proved.

4.2.4 Conditions for asymptotic stability

Assume that for the vector functions $f(t,x)$ and $g(t,x)$ from the system (4.2.1) the conditions of Lemma 4.2.3 are satisfied, that is, the functions $f(t,x)$ and $g(t,x)$ in the domain Ω satisfy the Lipshitz condition with respect to x with a constant L and $f(t,0) = g(t,0) = 0$.

Denote

$$\varkappa(t,t_0,x_0) = \frac{1}{t-t_0} \int_{t_0}^{t} \varphi(t,\overline{x}(t,t_0,x_0))\,dt - \Theta_0(t_0,x_0).$$

Theorem 4.2.3 *For the system (4.2.1) let the following conditions be satisfied:*

(1) *there exists a positive definite decrescent function $v(t,x)$ such that*

$$\dot{v}(t,x)|_{(4.2.3)} \leq v^*(x) \leq 0;$$

(2) *the function $\varphi(t,x)$ is differentiable with respect to x and there exist constants $M > 0$ and $d \geq 1$ such that the following inequalities hold:*

$$|\varphi(t,x)| \leq M\|x\|^d, \quad \|\nabla\varphi\| \leq M\|x\|^{d-1};$$

(3) *uniformly with respect to (t_0,x_0) there exists a mean $\Theta_0(t_0,x_0)$ and the following condition is satisfied:*

$$\max[\rho(x_0, E(v^* = 0)), (-\Theta_0(t_0,x_0))] > c_1(\|x_0\|), \quad c_1(r) \in K;$$

(4) *there exists a constant $k > 0$ such that for $x_0 \in \Omega$ the following condition is satisfied:*
$$\rho(x_0, E(v^* = 0)) < c_1(\|x_0\|),$$
and for $t_0 \in J$ the following inequality holds:
$$k\Theta_0(t_0,x_0) \leq -\|x_0\|^d;$$

(5) *there exist a constant $\mu_1 > 0$ and a function $c_2(r) \in K$ such that at x satisfying the condition $\rho(x, E(v^* = 0)) > c_1(\|x\|)$, and $\mu < \mu_1$, $t \in J$*

$$\dot{v}(t,x)|_{(4.2.3)} + \mu\varphi(t,x) < -c_2(\|x\|);$$

(6) *there exists a constant $l > 0$ such that at $t - t_0 > l$, $t_0 \in R_+$, $\rho(x_0, E(v^* = 0)) \leq c_1(\|x_0\|)$ the function $k(t,t_0,x_0)$ satisfies the estimate*

$$|\varkappa(t,t_0,x_0)| \leq |\Theta_0(t_0,x_0)|/4.$$

Then the solution $x = 0$ of the system (4.2.1) is asymptotically μ-stable.

Proof Let $\varepsilon \in (0, H)$, $t_0 \in R_+$, and $\alpha \in (0, \varepsilon)$ be specified. Show that there exist numbers $\eta(\varepsilon) > 0$, $\mu_0(\varepsilon) > 0$, $T(\mu, \alpha, \varepsilon) > 0$, not depending on t_0, such that for the perturbed motion $x(t, \mu)$ of the system (4.2.1) under the condition $\|x_0\| < \eta$ the following inequalities hold: $\|x(t, \mu)\| < \varepsilon$ for all $t \geq t_0$ and $\|x(t, \mu)\| < \alpha$ for $t > t_0 + T(\mu, \alpha, \varepsilon)$.

For the function $v(t, x)$ indicated in condition (1) of Theorem 4.2.3 the following inequality holds:

$$a(\|x\|) \leq v(t, x) \leq b(\|x\|), \qquad (4.2.38)$$

where $a(r) \in K$ and $b(r) \in K$.

Choose $\eta(\varepsilon) = b^{-1}(a(\varepsilon/2))$. In view of the inequality (4.2.38) for all points of the moving surface $v(t, x) = a(\varepsilon/2)$ the condition

$$\eta(\varepsilon) \leq \|x\| \leq \varepsilon/2$$

is satisfied, and all points of the surface $v(t, x) = a(\alpha/2)$ satisfy the inequality

$$b^{-1}(a(\alpha/2)) \leq \|x\| \leq \alpha/2$$

for all $t \in R_+$.

Consider some properties of the perturbed motion $x(t, \mu)$ at $\|x_0\| < \eta(\varepsilon)$ and $t_0 \in R_+$.

(a) Let at a point of time τ the following conditions be satisfied:

$$\|x(\tau)\| \geq b^{-1}(a(\alpha/2)), \quad v(\tau, x(\tau)) < a(\varepsilon/2),$$
$$\rho(x(\tau), E(v^* = 0)) > c_1(\|x(\tau)\|).$$

Estimate the change of the function $v(t, x)$ along the perturbed motion $x(t) = x(t, \tau, x(\tau))$. In this case, in view of condition (5) of Theorem 4.2.3, as long as the condition

$$x(t) \in U = \{x \colon b^{-1}(a(\alpha/2)) \leq \|x\| \leq \varepsilon/2, \ \rho(x(t), E(v^* = 0)) > c_1(\|x(t)\|)\}$$

is satisfied, the full derivative of the function $v(t, x)$ in view of the system (4.2.1) at $\mu < \mu_1$ satisfies the inequality

$$\dot{v}(t, x)|_{(4.2.1)} \leq -c_2(b^{-1}(a(\alpha/2))). \qquad (4.2.39)$$

The function $v(t, x(t))$ is not increscent, which means that the integral curve $x(t, \tau, x(\tau))$ will not cross the surface

$$v(t, x) = a(\varepsilon/2)$$

and will not leave the domain $\|x\| \leq \varepsilon/2$, at least until the condition

$$\rho(x(t), E(v^* = 0)) > c_1(\|x(t)\|).$$

is violated.

The perturbed motion $x(t)$ cannot permanently stay in the set U within a time interval equal to

$$l_1 = b(\varepsilon/2)c_2(b^{-1}(a(\alpha/2))).$$

Indeed, assuming that within a time interval equal to l_1, $x(t) \in U$, from the inequality (4.2.39) for $t = \tau + l_1$ obtain

$$v(t, x(t)) \leq v(\tau, x(\tau)) - c_2(b^{-1}(a(\alpha/2)))l < 0.$$

This contradicts the condition of positive definiteness of the function $v(t, x)$. The contradiction implies the existence of a point of time from the interval $(\tau, \tau + l_1)$, at which one of the following conditions will be violated:

$$b^{-1}(a(\alpha/2)) \leq \|x\|, \quad \rho(x, E(v^* = 0)) > c_1(\|x\|). \tag{4.2.40}$$

(b) Let the following conditions be satisfied at a point τ:

$$\|x(\tau)\| \geq b^{-1}(a(\alpha/2)), \quad v(\tau, x(\tau)) < a(\varepsilon/2),$$
$$\rho(x(\tau), E(v^* = 0)) \leq c_1(\|x(\tau)\|).$$

By condition (4) of Theorem 4.2.3, in this event

$$\Theta_0(\tau, x(\tau)) < -c_1(\|x(\tau)\|) \leq -c_1(b^{-1}(a(\alpha/2))). \tag{4.2.41}$$

Estimate the change of the function $v(t, x)$ along the perturbed motion $x(t, \mu)$. On the basis of Lemma 4.2.3 the inequalities (4.2.7) hold for the solutions $x(t, \mu)$ and $x(t)$ of systems (4.2.1) and (4.2.3). Using condition (2) of the theorem and the inequalities (4.2.7), obtain

$$|\varphi(t, x(t)) - \varphi(t, \overline{x}(t))| \leq \max \|\nabla \phi\| \|x(t) - \overline{x}(t)\|$$
$$\leq M\|x(\tau)\|^{d-1}(Q_x(t-\tau) + Q_{\overline{x}}(t-\tau))^{d-1}\mu\|x(\tau)\|Q(t-\tau)$$
$$= \mu\|x(\tau)\|^d B(t-\tau).$$

Using condition (4) of the theorem, obtain

$$|\varphi(t, x(t)) - \varphi(t, \overline{x}(t))| \leq \mu k B(t - \tau)|\overline{\varphi}(\tau, x(\tau))|. \tag{4.2.42}$$

Determine μ_2 from the inequalities

$$\mu_2 k B(2l + l_1) \leq \frac{1}{4}, \quad \mu_2 Q(2l + l_1) < 1. \tag{4.2.43}$$

The second inequality of (4.2.43) means that the perturbed motion $x(t, \mu)$ will not leave the domain $\|x\| < \varepsilon$ at $\mu < \mu_2$ at least on the interval $(2l + l_1)$.

Using the inequality (4.2.42) and the existence of the mean $\Theta_0(\tau, x(\tau))$, obtain an estimate for $v(t, x, (\tau))$:

$$v(t, x(t)) \leq v(\tau, x(\tau)) + \int_{\tau}^{t} \mu k B(t - \tau)|v(\tau, x(\tau))|\, dt$$

$$+ \mu(t - \tau)[\Theta_0(\tau, x(\tau)) + \varkappa(t, \tau, x(\tau))].$$

Taking into account the existence of the interval l [condition (6) of Theorem 4.2.3] and the inequalities (4.2.5) and (4.2.7), at $\mu < \mu_2$, $t \in [\tau + 2l, \tau + 2l + l_1]$ obtain

$$v(t, x(t)) \leq v(\tau, x(\tau)) - \mu l c_1(b^{-1}(a(\alpha/2))). \tag{4.2.44}$$

The perturbed motion $x(t, \mu)$ may leave the domain bounded by the surface $v(t, x) = a(\varepsilon/2)$, but due to the choice $\mu < \mu_2$ it will remain in the domain $\|x\| < \varepsilon$ within a time interval $2l + l_1$ and, as it is clear from the inequality (4.2.43), at some point of time from the interval $(\tau, \tau + 2l)$ it will return to the domain $v(t, x) < a(\varepsilon/2)$.

Properties (a) and (b) imply that if $\mu < \mu_0 = \min(\mu_1, \mu_2)$, $\|x_0\| < \eta$ and $t_0 \in R_+$, then the perturbed motion $x(t, \mu)$ will not leave the domain $\|x\| < \varepsilon$ at all $t \geq t_0$. The number $\eta(\varepsilon)$ was chosen without regard to t_0. Prove that at some point of time the perturbed motion will get into the domain $\|x\| < b^{-1}(a(\alpha/2))$.

Consider the sequence of points of time $t_i = t_0 + i(2l + l_1)$, $i = 1, 2, \ldots$. On each interval $[t_i, t_{i+1}]$ the function $v(t, x(t))$ in view of properties (a) and (b) decreases along the perturbed motion at least by $\mu l c_1(b^{-1}(a(\alpha/2)))$, $\mu \in (0, \mu_0)$. Indeed, if condition (a) holds at the initial point of the interval, then on the time interval l_1 there exists a point of time when one of the inequalities (4.2.40) is violated. Assume that the second inequality is violated. Then, taking that point of time as the initial one, using property (b), obtain the estimate (4.2.44) in the finite point of the interval. If property (b) holds in the initial point of the interval, we immediately obtain the estimate (4.2.44) for the finite point of the interval.

Let n be the smallest integer satisfying the condition

$$n \geq b(\varepsilon/2)/\mu l c_1(b^{-1}(a(\alpha/2))).$$

Assume that on the interval $[t_0, t_n]$, $t_n = t_0 + n(2l + l_1)$, the perturbed motion is in the domain $\|x\| \geq b^{-1}(a(\alpha/2))$. Then at a point of time t_n

$$v(t_n, x(t_n)) \leq v(t_0, x_0) - n\mu l c_1(b^{-1}(a(\alpha/2))) < 0$$

holds, which contradicts the condition of the positive definiteness of the function $v(t, x)$. The contradiction implies the existence of a point of time $t_1 \in (t_0, t_0 + n(2l + l_1))$ at which the condition $\|x(t_1, \mu)\| < b^{-1}(a(\alpha/2))$ is satisfied.

In the same way as it was proved that the perturbed motion $x(t, \mu)$ remained in the domain $\|x\| < \varepsilon$ at $\|x_0\| < \eta(\varepsilon)$ and $t \geq t_0$, it is possible to show that the perturbed motion $x(t, t_1.x(t_1))$ will remain in the domain $\|x\| < \alpha$ at all $t \geq t_1$ and $\mu < \mu_0$.

This proves that at all $t > t_0 + T$, where

$$T = n[2l + b(\varepsilon/2)/c_2(b^{-1}(a(\alpha/2)))],$$

the condition $\|x(t, t_0, x_0)\| < \alpha$ is satisfied. The number T was chosen without regard to t_0.

Theorem 4.2.3 is proved.

4.3 Stability on a Finite Time Interval

Let the trivial solution of the system (4.2.3) be stable, which is determined by the existence of a positive definite continuously differentiable Lyapunov function $v_0(t, x)$, whose derivative in view of the equations (4.2.3) is identically zero, that is, this is a critical case.

In this section the systems (4.2.1) are considered, whose mean (4.2.4) is alternating in any arbitrarily small neighborhood of zero.

Introduce the notation

$$\begin{aligned}
E_\Theta^+(t_0) &= \{(t_0, x_0) \in \Omega \colon \Theta_0(t_0, x_0) > 0\}, \\
E_\Theta^-(t_0) &= \{(t_0, x_0) \in \Omega \colon \Theta_0(t_0, x_0) < 0\}, \\
E_\Theta^\delta(t_0) &= \{(t_0, x_0) \in \Omega \colon \Theta_0(t_0, x_0) < -\delta\}
\end{aligned} \tag{4.3.1}$$

and show the conditions for the μ-stability on a finite interval.

Consider the following result.

Theorem 4.3.1 *Let the following conditions be satisfied:*

(1) *there exists a positive definite Lyapunov function $v_0(t, x)$ of the system (4.2.3), decrescent in the domain Ω;*

(2) *the full derivative of the function $v_0(t, x)$ along the paths of the system (4.2.3) is identically zero in the domain Ω;*

(3) *for any $\varepsilon, \rho > 0$, $\rho < \varepsilon$, there exists $\varphi_0 > 0$, $\varphi_0 = \text{const}$, such that $|\varphi(t, x)| < \varphi_0$ at all $t \in J$, as soon as $\rho < \|x\| < \varepsilon$.*

Then one can show $\sigma(\varepsilon)$, $\eta(\varepsilon)$, $T = \sigma(\varphi_0 \mu)^{-1}$ such that all solutions that satisfy the inequality $\|x_0\| < \eta$ at the initial point of time, for all $t \in [t_0, t_0 + T]$ do not leave the domain $\|x\| < \varepsilon$.

The next theorem enables us to improve the estimate of the time interval on which $\|x\| < \varepsilon$, if it is known that for a specified t_0 the value x_0 belongs to $E_{\ominus}^{-}(t_0)$, where $E_{\ominus}^{-}(t_0)$ is determined by the expression (4.3.1).

Theorem 4.3.2 *Let the following conditions be satisfied:*

(1) *there exists a positive definite Lyapunov function $v_0(t,x)$ of the system (4.2.3), decrescent in the domain Ω;*

(2) *the full derivative of the function $v_0(t,x)$ along the paths of the system (4.2.3) is identically zero in the domain Ω;*

(3) *uniformly with respect to $(t_0, x_0) \in \Omega$ there exists a mean $\Theta_0(t_0, x_0)$ alternating in the domain Ω and at a specified t_0 and all $\gamma > 0$, $\eta > 0$, $\gamma < \eta < H$, the following condition is satisfied:*

$$(B_\eta \setminus B_\gamma) \cap E_{\ominus}^{-}(t_0) \neq \varnothing,$$
$$B_\eta = \{x\colon \|x\| < \eta\}, \quad B_\gamma = \{x\colon \|x\| < \gamma\};$$

(4) *there exist a summable function $F(t)$, constants F_0, $\varphi_0(\bar{\varepsilon})$, $\varphi_0(\varepsilon)$, and M_0, and a nondecrescent function $\chi(\alpha)$, $\lim\limits_{\alpha \to 0} \chi(\alpha) = 0$ such that in the domain Ω*

$$|\varphi(t, x') - \varphi(t, x'')| < \chi(\|x' - x''\|)F(t),$$

$$\int_{t_1}^{t_2} F(t)\, dt \leq F_0(t_2 - t_1) \quad \text{at all } \tau < t_1 < t_2 < \infty,$$

$$\|g(t,x)\| < M_0;$$

at all $\rho > 0$, $\bar{\varepsilon} > 0$, $\rho < \bar{\varepsilon} < \varepsilon$, $|\varphi(t,x)| < \varphi_0(\bar{\varepsilon})$ at all $t \in J$, $\rho < \|x\| < \varepsilon$,

$$|\varphi(t,x)| < \varphi_0(\varepsilon) \quad \text{at all } \bar{\varepsilon} < \|x\| < \varepsilon.$$

Then for any $\varepsilon > 0$ and $\bar{\varepsilon} < \varepsilon$ one can find η, w_ε, $w_{\bar{\varepsilon}}$, μ_0, l, $T = 2l + l\delta[\varphi_0(\bar{\varepsilon})]^{-1} + (w_\varepsilon - w_{\bar{\varepsilon}})[\mu\varphi_0(\varepsilon)]^{-1}$ such that for all $t \in [t_0, t_0 + T]$ $\|x(t)\| < \varepsilon$ holds, if only this path begins in the domain where the mean $\Theta_0(t_0, x_0) < 0$ and $\|x_0\| < \eta$.

Proof Let $\varepsilon > 0$ ($\varepsilon < H$) be specified. Specify $\bar{\varepsilon} > 0$, $\bar{\varepsilon} = \text{const}$, so that $\bar{\varepsilon} < \varepsilon$. According to condition (1), there exists a positive definite function $w(x)$ such that

$$v_0(t,x) \geq w(x) \quad \text{at all } (t,x) \in \Omega. \tag{4.3.2}$$

Introduce $w_{\bar{\varepsilon}}$ so that the surface $S_{\bar{\varepsilon}} = \{x \in \Omega\colon w(x) = w_{\bar{\varepsilon}}\}$ would lie within the $\bar{\varepsilon}$-neighborhood of the origin of coordinates. The moving surface $S_v(t) = \{x \in \Omega\colon v_0(t,x) = w_{\bar{\varepsilon}}\}$ in view of (4.3.2) lies inside the surface $S_{\bar{\varepsilon}}$, hence $S_v(t) \subset B_{\bar{\varepsilon}}$.

Since $v_0(t,x)$ is decrescent in the domain Ω, one can find $\eta > 0$ such that B_η would lie inside the moving surface $S_v(t)$. Since $\Theta_0(t_0, x_0) < 0$, then there exists $\delta > 0$: $\Theta_0(t_0, x_0) < -\delta$. According to condition (3) of the theorem, the values δ and $\gamma > 0$ may be chosen so that $x_0 \in (B_\eta \setminus B_\gamma) \cap E_\Theta^\delta(t_0)$. Denote $c = v_0(t_0, x_0)$. Obviously, $c < w_{\bar\varepsilon}$. Let the point (t_0, x_0) lie on the moving surface $S_v'(t) = \{x \in \Omega: v_0(t,x) = c\}$. This surface is located inside the surface $S_v(t)$.

Consider the behavior of the function $v_0(t,x)$ along the solution $x = x(t; t_0, x_0)$ of the system (4.2.1). It is easy to show that one can find sufficiently large l and small μ_0 such that at $\mu < \mu_0$ on the interval $t \in [t_0, t_0 + 2l]$ the solution $x(t)$ belongs to B_ε and at $t \in [t_0 + l, t_0 + 2l]$ the estimate $v_0(t,x) < c - (t - t_0)\mu\delta/2$ holds.

Assuming $t_1 = t_0 + 2l$, $x_1 = x(t)$, obtain $v_0(t_1, x_1) < c - \mu l\delta$. Here the point x_1 lies inside the surface $S_v'(t)$. Denote $c_1 = v_0(t_1, x_1)$ and construct a surface $S_v''(t) = \{x \in \Omega: v_0(t,x) = c_1\}$.

The moving surface $S_v''(t)$ lies inside a fixed surface $S_{\bar\varepsilon}$. Consider the behavior of $v_0(t,x)$ along a portion of the path of the system (4.2.1), beginning in the point (t_1, x_1). In view of the system (4.2.1) $dv_0/dt = \mu\varphi(t,x)$, hence, integrating from the point t_1, obtain

$$v_0(t_1 + T, x) = c_1 + \mu \int_{t_1}^{t_1+T} \varphi(t,x)\,dt < c - \mu l\delta + \mu\varphi_0 T < w_{\bar\varepsilon} - \mu l\delta + \mu\varphi_0 T.$$

For the path $x = x(t)$ to remain within $B_{\bar\varepsilon}$ on the time interval $t \in [t_1, t_1 + T_1]$, it suffices that the equality $w_{\bar\varepsilon} - \mu l\delta + \mu\varphi_0(\bar\varepsilon)T_1 = w_{\bar\varepsilon}$ should hold, whence $T_1 = l\delta[\varphi_0(\bar\varepsilon)]^{-1}$. Here, when estimating the function $\varphi(t,x)$ it is taken into account that x belongs to $B_{\bar\varepsilon}$.

Introduce the quantity w_ε so that the surface $S_\varepsilon = \{x \in \Omega: w(x) = w_\varepsilon\}$ should lie inside the ε-neighborhood of the origin of coordinates, $S_\varepsilon \subset B_\varepsilon$. The path $x = x(t, \mu)$ may only leave the ε-neighborhood of the origin of coordinates after crossing the moving surface $S_v''(t) = \{x \in \Omega: v_0(t,x) = w_\varepsilon\}$, which lies inside the surface S_ε. Therefore, for the path $x = x(t, \mu)$ to remain within the bounds of B_ε on the interval $t \in [t_1+T_1, t_1+T_1+T_2]$, it suffices that the equality $w_{\bar\varepsilon} + \mu\varphi_0 T_2 = w_\varepsilon$ should hold, whence $T_2 = (w_\varepsilon - w_{\bar\varepsilon})[\mu\varphi_0(\varepsilon)]^{-1}$. Here the function $\varphi(t,x)$ is estimated at $\rho < \|x\| < \varepsilon$.

Thus, on the whole interval $[t_0, t_0 + T]$, where

$$T = t_1 + T_1 + T_2 = 2l + \frac{l\delta}{\varphi_0(\bar\varepsilon)} + \frac{w_\varepsilon - w_{\bar\varepsilon}}{\mu\varphi_0(\varepsilon)},$$

the solution $x(t, \mu)$ with the initial value $x_0 \in E_\Theta^\delta(t_0) \cap B_\eta$ will not leave the ε-neighborhood of the stationary point.

The theorem is proved.

The next theorem allows us to essentially improve the estimate from Theorem 4.3.2 due to fuller information on the change of weak perturbations in the neighborhood of the equilibrium.

Theorem 4.3.3 *Let the following conditions be satisfied:*

(1) *there exists a positive definite Lyapunov function $v_0(t,x)$ of the system (4.2.3), decrescent in the domain Ω;*

(2) *the full derivative of the function $v_0(t,x)$ along the paths of the system (4.2.3) is identically zero in the domain Ω;*

(3) *for any ε, ρ such that $0 < \rho < \varepsilon < H$, in the domain $\rho < \|x\| < \varepsilon$ the function $\varphi(t,x)$ is defined and on the interval (ρ, ε) there exists a nondecrescent function $\psi(\zeta)$ such that $|\varphi(t,x)| \leq \psi(\zeta)$ at $\rho < \|x\| < \zeta$.*

Then on the interval (ρ, ε) there exists a nondecrescent continuous function $\varkappa(\zeta)$ and for any $\varepsilon_0 \in (\rho, \varepsilon)$ one can find $\eta(\varepsilon_0)$ such that all solutions satisfying the inequality $\|x_0\| < \eta$ at the initial point of time do not leave the domain $\|x\| < \varepsilon$ for all $t \in [t_0, t_0 + T]$, where

$$T = \frac{1}{\mu} \int_{\varepsilon_0}^{\varepsilon} \frac{d\varkappa(\zeta)}{\psi(\zeta)} \, . \tag{4.3.3}$$

Proof Let $\varepsilon > 0$ ($\varepsilon < H$) be specified. Specify $\varepsilon_0 > 0$, $\varepsilon_0 = \text{const}$, so that $\rho < \varepsilon_0 < \varepsilon$. Divide the interval $[\varepsilon_0, \varepsilon]$ by the points ε_i, $\varepsilon_0 < \varepsilon_1 < \ldots < \varepsilon_{k-1} < \varepsilon_k = \varepsilon$, and introduce the system of nested neighborhoods $\{B_i\}$, $B_i = \{x \colon \|x\| < \varepsilon_i\}$.

According to condition (1), there exists a positive definite function $w(x)$ such that

$$v_0(t,x) \geq w(x) \quad \text{at all} \quad (t,x) \in \Omega. \tag{4.3.4}$$

On the interval $[\varepsilon_0, \varepsilon]$ introduce a continuous nondecrescent function $\varkappa(\zeta)$ so that at any $\zeta \in [\varepsilon_0, \varepsilon]$ the surface $\{x \in \Omega \colon w(x) = \varkappa(\zeta)\}$ would lie in the neighborhood B_ζ, and construct the system of surfaces $S_i = \{x \in \Omega \colon w(x) = \varkappa(\varepsilon_i)\} \subset B_i$, $i = 0, 1, 2, \ldots, k$. Since the function $w(x)$ is continuous and $\varkappa(\zeta)$ nondecrescent, all the surfaces S_i are closed and the surface S_i lies inside the surface S_{i+1}. Consider the system of moving surfaces $\{\overline{S}_i\}$, $\overline{S}_i = \{x \in \Omega \colon v_0(t,x) = \varkappa(\varepsilon_i)\}$. In view of the inequality (4.3.4) the surface $\overline{S}_i$ lies inside the surface S_i. The continuity of $v_0(t,x)$ implies that the surfaces $\overline{S}_i$ are closed; in addition, the surface $\overline{S}_i$ lies inside the surface $\overline{S}_{i+1}$. Therefore, the solution may only leave the neighborhood B_i after crossing $\overline{S}_i$.

Since $v_0(t,x)$ is decrescent in the domain Ω, one can find $\eta > 0$ such that the η-neighborhood B_η of the point $x = 0$ for all $t > 0$ lies within the moving surface $\overline{S}_0$.

Let the integral curve $x(t; t_0, x_0)$ leave the neighborhood B_η and assume that at a point of time $t = t_0$ it crossed the surface $\overline{S}_0$. Let t_i denote the point of time when the curve crosses the surface $\overline{S}_i$, and let x_i denote the respective value of x: $x_i = x(t_i; t_0, x_0)$.

Differentiate $v_0(t, x)$ in view of the equations (4.2.1). Taking into account conditions (2) and (3) of Theorem 4.3.3, obtain

$$\frac{dv_0}{dt} = \mu\varphi(t, x). \tag{4.3.5}$$

Consider the behavior of the function $v_0(t, x)$ along the portion of the path, beginning at the point (t_i, x_i). Integrating (4.3.5) from the point $t = t_i$, in view of condition (3) and the definition of t_i obtain

$$v_0(t_{i+1}, x_{i+1}) = v_0(t_i, x_i) + \mu \int_{t_i}^{t_{i+1}} \varphi(t, x)\, dt \leq v_0(t_i, x_i) + \mu\psi(\varepsilon_{i+1})(t_{i+1} - t_i).$$

The last inequality, taking into account that $v_0(t_i, x_i) = \varkappa(\varepsilon_i)$, implies

$$t_{i+1} - t_i \geq \frac{\varkappa(\varepsilon_{i+1}) - \varkappa(\varepsilon_i)}{\mu\psi(\varepsilon_{i+1})}, \qquad i = 0, 1, \ldots, k-1,$$

and

$$t_k - t_0 \geq \frac{1}{\mu} \sum_{i=1}^{k-1} \frac{\varkappa(\varepsilon_{i+1}) - \varkappa(\varepsilon_i)}{\psi(\varepsilon_{i+1})}.$$

In view of the definition of the function $\varkappa(\zeta)$ and the arbitrariness of the division of the interval $[\varepsilon_0, \varepsilon]$, the sum in the right-hand part is integral one. Passing to the limit at $\max(\varepsilon_i - \varepsilon_{i-1}) \to 0$, obtain

$$T = \frac{1}{\mu} \int_{\varepsilon_0}^{\varepsilon} \frac{d\varkappa(\zeta)}{\psi(\zeta)}.$$

The theorem is proved.

For solutions beginning in the domain where $\Theta_0(t_0, x_0) < 0$, the following theorem is correct.

Theorem 4.3.4 *Let:*

(1) *all the conditions of Theorem 4.3.2 be satisfied;*

(2) *uniformly with respect to $(t_0, x_0) \in \Omega$ there exists a mean $\Theta_0(t_0, x_0)$, which is alternating sign in the domain Ω, besides at the specified t_0 and at all $\gamma > 0$, $\eta > 0$, $\gamma < \eta < H$, the following condition is satisfied:*

$$(B_\eta \setminus B_\gamma) \cap E_\Theta^-(t_0) \neq \varnothing,$$
$$B_\eta = \{x \colon \|x\| < \eta\}, \qquad B_\gamma = \{x \colon \|x\| < \gamma\};$$

(3) *there exists a summable function $F(t)$, constants F_0, M_0, and a nondecrescent function $\chi(\alpha)$, $\lim_{\alpha \to 0} \chi(\alpha) = 0$, such that in the domain Ω*

$$|\varphi(t, x') - \varphi(t, x'')| < \chi(\|x' - x''\|)F(t), \quad \|g(t, x)\| < M_0,$$

$$\int_{t_1}^{t_2} F(t)\, dt \leq F_0(t_2 - t_1) \quad at\ all\ \ \tau < t_1 < t_2 < \infty.$$

*Then for any $\varepsilon > 0$ and $\rho < \bar{\varepsilon} < \varepsilon$ on the interval $[\rho, \varepsilon]$ there exists a
nondecrescent continuous function $\varkappa(\zeta)$ and one can find $\eta(\bar{\varepsilon})$, μ_0, l, $\varepsilon_0 < \bar{\varepsilon}$,*

$$T = 2l + \frac{1}{\mu} \int_{\varepsilon_0}^{\varepsilon} \frac{d\varkappa(\zeta)}{\psi(\zeta)} \tag{4.3.6}$$

*such that for all $t \in [t_0, t_0 + T]$ $\|x(t)\| < \varepsilon$ holds, if only this path begins in
the domain where the mean $\Theta_0(t_0, x_0) < 0$ and $\|x_0\| < \eta$.*

Proof Let $\varepsilon > 0$ ($\varepsilon < H$) be specified. Specify $\bar{\varepsilon} > 0$, $\bar{\varepsilon} = $ const so that
$\bar{\varepsilon} < \varepsilon$. Repeating the reasoning from the proof of Theorem 4.3.3, introduce
a continuous and nondecrescent function $\varkappa(\zeta)$ on the interval $[\rho, \varepsilon]$. In view
of condition (1) of the theorem, $\overline{S}_\varepsilon = \{x \in \Omega : v_0(t, x) = \varkappa(\bar{\varepsilon})\}$ lies in the
neighborhood $B_{\bar{\varepsilon}}$.

Since $v_0(t, x)$ is decrescent in the domain Ω, one can find $\eta > 0$ such that
B_η lies inside the moving surface $\overline{S}_{\bar{\varepsilon}}$. Like in the proof of Theorem 4.3.2, verify
that one can find sufficiently large l and small μ_0 such that at $\mu < \mu_0$ the
solution $x(t)$, leaving the point $x_0 \in (B_\eta \setminus B_\gamma) \cap E_\Theta^\delta(t_0)$ at a point of time
t_0, on the interval $t \in [t_0, t_0 + 2l]$ does not leave B_ε and at $t = t_1 = t_0 + 2l$
$v_0(t_1, x_1) < v_0(t_0, x_0) - l\delta\mu$ holds, where $x_1 = x(t_1)$ and $v_0(t_0, x_0) < \varkappa(\bar{\varepsilon})$.

Let ε_0 be a root of the equation $\varkappa(\varepsilon_0) = \varkappa(\bar{\varepsilon}) - l\delta\mu$. Construct the surfaces
$\overline{S}_{\varepsilon_0} = \{x \in \Omega : v_0(t, x) = \varkappa(\varepsilon_0)\}$ and $\overline{S}_\varepsilon = \{x \in \Omega : v_0(t, x) = \varkappa(\varepsilon)\}$. By
the definition of the function $\varkappa(\zeta)$ the surfaces $\overline{S}_{\varepsilon_0}$ and $\overline{S}_\varepsilon$ are closed and
$\overline{S}_{\varepsilon_0} \subset B_{\varepsilon_0}$, $\overline{S}_\varepsilon \subset B_\varepsilon$. Since the point x_1 is located inside the moving surface
$\overline{S}_{\varepsilon_0}$, an integral curve emanating from it may only leave B_ε after sequential
crossing of the surfaces $\overline{S}_{\varepsilon_0}$ and $\overline{S}_\varepsilon$.

Assume that the integral curve crossed the surface $\overline{S}_{\varepsilon_0}$ at a point of time
$t = t_2 > t_1$. Repeating the reasoning of the proof of Theorem 4.3.3, verify that
the solution will not leave the domain $\|x\| < \varepsilon$ for all $t \in [t_2, t_2 + T_1]$, where

$$T_1 = \frac{1}{\mu} \int_{\varepsilon_0}^{\varepsilon} \frac{d\varkappa(\zeta)}{\psi(\zeta)}.$$

Thus, on the whole interval $[t_0, t_0 + T]$, where

$$T = t_1 + T_1 - t_0 = 2l + \frac{1}{\mu} \int_{\varepsilon_0}^{\varepsilon} \frac{d\varkappa(\zeta)}{\psi(\zeta)},$$

the solution $x(t)$ with the initial value $x_0 \in e_\Theta^\delta(t_0) \cap B_\eta$ will not leave the ε-neighborhood of the point $x = 0$.

The theorem is proved.

4.4   Methods of Application of Auxiliary Systems

The results of the previous section are based on the known general solution of the shortened system (4.2.3), which essentially confines the area of application of such approach in the study of stability or instability of motion. We will show that instead of solutions of the system (4.2.3) it is possible to use solutions of some limiting system, which may turn out to be simpler.

4.4.1   Development of limiting system method

Consider the systems (4.2.1) and (4.2.3) under the same assumptions with regard to the right-hand part.

Let for some system

$$\frac{dx}{dt} = f^0(t, x), \quad x(t_0) = x_0, \tag{4.4.1}$$

connected with the system (4.2.3), the general solution $x^0(t) = x^0(t, t_0, x_0) \in D_1$ is known at $(t_0, x_0) \in \operatorname{int} \Omega$ and $t \geq t_0$. The vector function $f^0(t, x)$ is continuous and satisfies the Lipschitz condition with respect to x with the constant L in the domain $R_+ \times D_1$.

Instead of the mean $\Theta_0(t_0, x_0)$ we will use an integral calculated along solutions of the system (4.4.1).

Denote

$$\dot{v}^0(t, x)\big|_{(4.4.1)} = \frac{\partial v^0}{\partial t} + \left(\frac{\partial v^0}{\partial x}\right)^{\mathrm{T}} f^0(t, x),$$

$$G^0(T, t_0, x_0) = \int\limits_{t_0}^{t_0+T} \varphi(t, x^0(t, t_0, x_0))\, dt.$$

Theorem 4.4.1 *Let for the system (4.2.1) the following conditions be satisfied:*

(1) *there exists a positive definite decrescent function $v(t, x)$ and*

$$\dot{v}(t, x)\big|_{(4.2.3)} \leq 0;$$

(2) *for any number $\xi \in (0, H)$ uniformly with respect to $x \in \{x : \|x\| < \xi\}$ there exists a limit*

$$\lim_{t \to \infty} \|f(t, x) - f^0(t, x)\| = 0;$$

(3) *there exist a $\chi(\beta) \in K$, summable functions $M(t)$ and $F(t)$ such that on any finite interval $[t_1, t_2]$, $t_1, t_2 \in R_+$, the following inequalities hold:*

$$\|g(t, x)\| \leq M(t), \quad |\varphi(t, x') - \varphi(t, x'')| \leq F(t)\chi(\|x' - x''\|)$$

and

$$\int_{t_1}^{t_2} M(t)\, dt \leq M_0(t_2 - t_1), \quad \int_{t_1}^{t_2} F(t)\, dt \leq F_0(t_2 - t_1),$$

$$M_0, F_0 = \text{const};$$

(4) *for any numbers α, β, $0 < \alpha < \beta < H$, there exist positive quantities μ', δ, l such that for the values t', x', $t' \in R_+$, $\alpha \leq \|x'\| \leq \beta$, one of the following conditions is satisfied:*

   (a) $\dot{v}(t', x')\big|_{(4.2.3)} + \mu\varphi(t', x') \leq 0$ *at* $\mu < \mu'$,

   (b) $G^0(T, t', x') \leq -\delta T$ *at* $T > l$.

Then the solution $x = 0$ of the system (4.2.1) is μ-stable.

Proof Let $\varepsilon \in (0, H)$ be satisfied. In view of condition (1) of Theorem 4.4.1, all points of the moving surface $v(t, x) = a(\varepsilon/2)$ satisfy the inequality (4.2.11) for all $t \in R_+$. Under condition (4) for the numbers $b^{-1}(a(\varepsilon/2))$ and $\varepsilon/2$, there exist positive constants δ and l. It can be shown that for

$$\lambda = \min\left[\chi^{-1}(\delta/4F_0)/2l\exp(2lL),\ \varepsilon/4l\exp(2lL)\right]$$

there exists a point of time τ_0 such that at $\tau > \tau_0$ and $t \in [\tau, \tau + 2l]$ for the solutions $\bar{x}(t, \tau, x_\tau)$ and $x^0(t, \tau, x_\tau)$ of the systems (4.2.3) and (4.4.1) the following inequality holds:

$$\|\bar{x}(t, \tau, x_\tau) - x^0(t, \tau, x_\tau)\| \leq \min\left[\chi^{-1}(\delta/4F_0),\ \varepsilon/2\right] \qquad (4.4.2)$$

at x_τ, satisfying the condition $v(\tau, x_\tau) = a(\varepsilon/2)$. Indeed, condition (2) implies that for λ one can find a point τ_0 such that

$$\|f(t, x) - f^0(t, x)\| < \lambda$$

at $t > \tau_0$, $\|x\| < \varepsilon$. From the equations (4.2.3) and (4.4.1) for $\tau > \tau_0$ and

$t \in [\tau, \tau + 2l]$ obtain

$$\|\overline{x}(t, \tau, x_\tau) - x^0(t, \tau, x_\tau)\| \leq \int_\tau^t \|f(t, \overline{x}(t, \tau, x_\tau)) - f^0(t, \overline{x}(t, \tau, x_\tau))\| \, dt$$

$$+ \int_\tau^t \|f^0(t, \overline{x}(t, \tau, x_\tau)) - f^0(t, x^0(t, \tau, x_\tau))\| \, dt$$

$$< \lambda 2l \int_\tau^t L \|\overline{x}(t, \tau, x_\tau) - x^0(t, \tau, x_\tau)\| \, dt.$$

Using Lemma 4.2.1, obtain the estimate (4.4.2). From the equation (4.2.1), taking into account condition (3) and the Lipschitz condition for $f(t, x)$, obtain

$$\|x(t, t_0, x_0)\| \leq \|x_0\| + \int_{t_0}^t \|f(t, x(t, t_0, x_0))\| \, dt + \mu \int_{t_0}^t M(t) \, dt$$

$$\leq \|x_0\| + \mu M_0 (t - t_0) + \int_{t_0}^t L \|x(t, t_0, x_0)\| \, dt.$$

Now, applying Lemma 4.2.1, for any $t_0 \in [0, \tau_1]$ and $t \in [t_0, \tau_1]$ (let $\tau_0 < \tau_1 < \infty$) obtain the estimate

$$\|x(t, t_0, x_0)\| \leq [\|x_0\| + \mu M_0 \tau_0] \exp(L\tau_1). \tag{4.4.3}$$

Choose

$$\mu_1 = b^{-1}(a(\varepsilon/2))/2M_0\tau_1 \exp(L\tau_1), \quad \eta = b^{-1}(a(\varepsilon/2))/2 \exp(L\tau_1).$$

Then the estimate (4.4.3) implies the estimate

$$\|x(t, t_0, x_0)\| < b^{-1}(a(\varepsilon/2))$$

at $\|x_0\| < \eta$, $t_0 \in [0, \tau_1]$, $t \in [t_0, \tau_1]$. Therefore, to prove the theorem it suffices to show that $\|x(t, t_0', x_0')\| < \varepsilon$ at $t_0' \geq \tau_1$ and $\|x_0'\| < b^{-1}(a(\varepsilon/2))$ at $t \geq t_0'$. Consider the solution $x(t, \mu)$ and assume that it has left the domain $\|x\| < b^{-1}(a(\varepsilon/2))$ and at some point of time $t = \tau$ the condition $v(\tau, x(\tau)) = a(\varepsilon/2)$ is satisfied. At a point $x(\tau)$ the inequality (4.2.11) will hold and one of the conditions (4a) or (4b) will be satisfied.

1. Let condition (4a) be satisfied at a point of time τ, that is, for the numbers $b^{-1}(a(\varepsilon/2))$ and $\varepsilon/2$ there exists μ' such that the full derivative of the Lyapunov function in view of the system (4.2.1) at that point is nonpositive at $\mu < \mu'$, which means that the solution $x(t)$ at the point τ cannot cross the surface

$$v(\tau, x(\tau)) = a(\varepsilon/2).$$

2. Let condition (4b) be satisfied at the point τ, that is, for the numbers $b^{-1}(a(\varepsilon/2))$ and $\varepsilon/2$ there exist δ and l such that

$$G^0(T, \tau, x(\tau)) \leq -\delta T \tag{4.4.4}$$

at $T > l$. Integrating the expression of full derivative of the Lyapunov function in view of the system (4.2.1), for $t > \tau$ obtain

$$v(t, x(t, \tau, x(\tau))) \leq v(\tau, x(\tau)) + \mu \int_{\tau}^{t} \varphi(t, x(t, \tau, x(\tau)))\, dt. \tag{4.4.5}$$

Represent the last integral in the form

$$
\begin{aligned}
\int_{\tau}^{t} \varphi(t, x(t))\, dt &= \int_{\tau}^{t} [\varphi(t, x(t)) - \varphi(t, \overline{x}(t))]\, dt \\
&+ \int_{\tau}^{t} [\varphi(t, \overline{x}(t)) - \varphi(t, x^0(t))]\, dt + \int_{\tau}^{t} \varphi(t, x^0(t))\, dt.
\end{aligned} \tag{4.4.6}
$$

Here $x(t) = x(t, \tau, x(\tau))$, $\overline{x}(t) = \overline{x}(t, \tau, x(\tau)))$, $x^0(t) = x^0(t, \tau, x(\tau))$. On the basis of Lemma 4.2.2 for the norm of difference of solutions $x(t)$ and $\overline{x}(t)$ at $t \in [\tau, \tau + 2l]$ the following estimate is true:

$$\|x(t) - \overline{x}(t)\| \leq \mu M_0 l \exp(2lL). \tag{4.4.7}$$

Choose

$$\mu_2 = \frac{\varepsilon}{4M_0 l \exp(2lL)}.$$

Then at $\mu < \mu_2$ from the inequality (4.4.7) obtain $\|x(t) - \overline{x}(t)\| < \varepsilon/2$ at all $t \in [\tau, \tau + 2l]$. In view of condition (1) of Theorem 4.4.1 and the inequality $\|\overline{x}(t)\| < \varepsilon/2$ obtain $\|x(t)\| \leq \varepsilon$ at $t \in [\tau, \tau + 2l]$. Estimate the first integral in the right-hand part of the expression (4.4.6). Choose

$$\mu_3 = \chi^{-1}(\delta/4F_0)/2M_0 l \exp(2lL).$$

Using condition (3) and the inequality (4.4.7), for $\mu < \min(\mu_2, \mu_3)$ and $t \in [\tau, \tau + 2l]$ obtain

$$\int_{\tau}^{t} |\varphi(t, x(t)) - \varphi(t, \overline{x}(t))|\, dt \leq \int_{\tau}^{t} F(t)\chi(\|x(t) - \overline{x}(t)\|)\, dt \leq \delta(t - \tau)/4. \tag{4.4.8}$$

From the inequality (4.4.2) it is clear that the solution $x^0(t, \mu)$ on the interval $[\tau, \tau + 2l]$ will not leave the ε-neighborhood and, respectively, the domain Ω; therefore, for the estimation of the second integral from the expression

(4.4.6) it is possible to use condition (2) of the theorem. Taking into account the inequality (4.4.2), for $t \in [\tau, \tau + 2l]$ obtain the estimate

$$\int_\tau^t |\varphi(t, \bar{x}(t)) - \varphi(t, x^0(t))| \, dt \leq \int_\tau^t F(t)\chi(\|\bar{x}(t) - x^0(t)\|) \, dt \leq \delta(t - \tau)/4.$$

$$(4.4.9)$$

The expression (4.4.6) and the inequalities (4.4.4), (4.4.8), and (4.4.9) at $\mu < \min(\mu_2, \mu_3)$ and $t \in [\tau + l, \tau + 2l]$ imply the estimate

$$\int_\tau^t \varphi(t, x(t)) \, dt \leq \delta(t - \tau)/4.$$

Thus, the integral in the inequality (4.4.5) becomes negative at least from the point $t = \tau + l$, which means that the solution $x(t, \mu)$, having left the surface $v(\tau, x(\tau)) = a(\varepsilon/2)$, due to the choice of μ will remain in the domain $\|x\| < \varepsilon$ at $t \in [\tau, \tau + 2l]$ and at some point of time from the interval $[\tau, \tau + 2l]$ will return to the domain bounded by the surface $v(\tau, x(\tau)) = a(\varepsilon/2)$.

From the considered cases 1 and 2 and the choice of η, it is clear that the solution $x(t, \mu)$ at $t_0 \in R_+$, $\|x_0\| < \eta$, $\mu < \mu_0 = \min(\mu_1, \mu_2, \mu_3, \mu')$ will not leave the domain $\|x\| < \varepsilon$ at all $t \geq t_0$.

The theorem is proved.

Let us use a simple example to illustrate Theorem 4.4.1.

Example 4.4.1 Study the equilibrium state $x = 0$ of the system

$$\frac{dx_1}{dt} = -x_1 + [1 - p(t)]x_2 + \mu(x_1^3 - ax_2^3),$$

$$\frac{dx_2}{dt} = p(t)(x_1 - x_2) + \mu[a(x_1 + x_2)\cos t + (x_2^3 - ax_1^3)].$$

$$(4.4.10)$$

Here $a = \text{const} > 1$, $p(t)$ is a continuous function, $0 \leq p(t) \leq m = \text{const}$, $\lim_{t \to \infty} p(t) = 0$.

The derivative of the Lyapunov function

$$v = [x_2^2 + (x_1 - x_2)^2]/2$$

along solutions of the system (4.4.10) at $\mu = 0$ has the form

$$\dot{v} = -(1 + p)(x_1 - x_2)^2 \leq 0.$$

The integral $G^0(T, t_0, x_0)$ calculated along the solutions

$$x_1^0 = x_{20} + (x_{10} - x_{20}) \exp[-(t - t_0)],$$
$$x_2^0 = x_{20}$$

of the limiting system $\dot{x}_1 = -x_1 + x_2$, $\dot{x}_2 = 0$, will satisfy condition (4b) of

Theorem 4.4.1 at $t_0 \geq 0$ and $x_0 \in E = \{x_0 \colon |x_{20}| \leq 2|x_{10}|, \ |x_{10}| \leq 2|x_{20}|\}$, since there exists a mean

$$\lim_{T \to \infty} \frac{1}{T} G^0(T, t_0, x_0) = x_{20}^4 (1 - a),$$

negative at $a > 1$ and at the values $t_0 \geq 0$ and $x_0 \in E$.

In the remaining part of the neighborhood of zero, in view of the negativeness of $\dot{v}(t, x)$, condition (4a) will be satisfied. Thus, the stability of the equilibrium state $x = 0$ of the system (4.4.10) holds in the following sense: for any ε there exists $\eta(\varepsilon) > 0$ and $\mu(\varepsilon) > 0$ such that $\|x(t, t_0, x_0)\| < \varepsilon$ at all $t \geq t_0$, as soon as $\|x_0\| < \eta$, $t_0 \geq 0$.

Similarly, it is possible to show that the theorem of instability is correct. We will give its statement without proof.

Theorem 4.4.2 *Let for some $\tau \in R_+$ in the domain $[\tau, \infty) \times D$:*

(1) *there exists a domain $v > 0$, in which the function $v(t, x)$ is bounded, and*

$$\dot{v}(t, x)|_{(4.2.3)} \geq 0, \quad \lim_{t \to \infty} \dot{\Theta}_0^0(t, x) \geq 0;$$

(2) *in the domain $v > 0$, conditions (2) and (3) of Theorem 4.4.1 be satisfied;*

(3) *for an arbitrarily small number $\alpha > 0$ there exist positive numbers μ', γ and δ, l such that for each value t', x', satisfying the inequality $v(t', x) > \alpha'$ one of the following conditions is satisfied:*

  (a) $\dot{v}(t', x')|_{(4.2.3)} + \mu \varphi(t', x') \geq \gamma$ *at* $\mu < \mu'$,
  (b) $G^0(T, t', x') \geq \delta T$ *at* $T > l$.

Then the solution $x = 0$ of the system (4.2.1) is μ-unstable.

Note that the derived Lyapunov function in view of the system (4.2.1) may be alternating in the domain $v > 0$ in contrast to the limitations on the function v in Chetaev's instability theorem [1].

4.4.2   Stability on time-dependent sets

Let $\mathcal{P}$ be the set of all subsets of the set R^n. The mapping $\mathbb{S} \colon R \to \mathcal{P}$ is a set-valued function. The set of its values at $t \in R$ is a time-varying set $\mathbb{S}(t)$. Let $\mathbb{N} = \{(t, x) \colon t \in R, \ x \in N(t)\}$, $N(t)$ be a time-varying neighborhood of the point $x = 0$. If $\mathbb{S}(t)$ is substituted with $N(t)$, then $\mathbb{S}$ is substituted with $\mathbb{N}$.

Now consider the system of the form

$$\frac{dx}{dt} = f(t, x) + g(t, x), \quad x(t_0) = x_0. \tag{4.4.11}$$

Along with the shortened system

$$\frac{dx}{dt} = f(t, x), \quad x(t_0) = x_0$$

consider the limiting system

$$\frac{dx}{dt} = f^0(\beta, x), \quad f^0(\beta, 0) \equiv 0. \tag{4.4.12}$$

Here it is assumed that at all $x \in \mathbb{S}(t)$ there exists a limit

$$\lim_{t \to \beta} \| f(t, x) - f^0(\beta, x) \| = 0, \quad 0 < \beta < \tau.$$

Remark 4.4.1 If $\beta = +\infty$ and the approach to the limit is uniform with respect to $x \in \{x : \|x\| < H\}$, then the limiting system (4.4.12) coincides with the one considered in Section 4.4.1.

Assume that the nonperturbed motion $x = 0$ of the system (4.4.12) is nonasymptotically stable and for it there exists a function $v(t, x)$ with the respective properties.

Theorem 4.4.3 *Let the vector function f^0 be continuous at $(t, x) \in \mathbb{N}$ and let $N(t)$ be a continuous neighborhood of the point $x = 0$ at each $t \in R$, which may be time-invariant or time-variable and the following conditions are satisfied:*

(1) *for the system (4.4.12) there exist a function $v(t, x)$ and functions a, b belonging to the K-class, such that at any $(t, x) \in \mathbb{S}$*

    (a) $a(\|x\|) \leq v(t, x) \leq b(\|x\|)$,

    (b) $dv/dt \leq 0$;

(2) *at any ρ, $0 < \rho < H < +\infty$, in the domain $\{x : \rho \leq \|x\| < H\} \subset N(t)$ the function $\varphi(t, x) = (\partial v/\partial x)^{\mathrm{T}} g(t, x)$ is defined and there exist a function c from the K-class and a continuous nonnegative function $\varkappa(t)$ at all $t \in J$ such that $|\varphi(t, x)| \leq c(\|x\|)\varkappa(t)$;*

(3) *there exist a function w from the K-class and a continuous nonnegative function $\lambda(t)$ such that*

$$|(\partial v/\partial x)^{\mathrm{T}}(f - f^0)| \leq w(\|x\|)\lambda(t) \quad \text{at all} \quad (t, x) \in \mathbb{S}.$$

Then for any $\varepsilon_0 > \rho$ there exists $\eta(\varepsilon_0)$ such that any motion of the system (4.4.11), which begins at $t = t_0$ in the domain $\|x_0\| \leq \eta(\varepsilon_0)$, will not leave the set $B_\zeta = \{x : \|x\| \leq \zeta(t)\}$ on the interval of existence of a positive solution of the differential equation

$$\frac{d\zeta}{dt} = \lambda(t) \frac{w(\zeta)}{a'(\zeta)} + \varkappa(t) \frac{c(\zeta)}{a'(\zeta)}, \quad \zeta(t_0) = \varepsilon_0 > 0, \tag{4.4.13}$$

such that $\zeta(t) < H$.

Proof Let $\varepsilon_0 \in (\rho, H)$. For the function $\zeta(t)$ with the value $\zeta \in [\varepsilon_0, H]$ at $t \in \mathcal{T} = [t_0, t_0 + \tau)$ (τ is a finite number or $+\infty$) consider a time-varying set $B_\zeta = \{x \colon \|x\| \leq \zeta(t)\}$ and a moving surface $S_\zeta = \partial\mathcal{V}_\zeta(t)$, $\mathcal{V}_\zeta(t) = \{x \colon v(t, x) < a(\zeta(t))\}$. From the continuity of $v(t, x)$ it follows that the surface S_ζ is closed. According to condition (1a) obtain $S_\zeta \subset B_\zeta$. Indeed, for $\|x\| = \zeta(t)$ at all $t \in J_1 \subset \mathcal{T}$ we have $v(t, x) \geq a(\zeta(t))$, and in view of the monotonicity of the function $a(\cdot)$ the equality $v(t, x) = a(\zeta)$ will hold at the points $x \in B_\zeta$. Hence it follows that $S_\zeta \subset B_\zeta$ and at any ζ_1, ζ_2 such that $\varepsilon_0 < \zeta_1 < \zeta_2 < H$, the surface S_{ζ_1} is embedded in S_{ζ_2}. Therefore, at any $\varepsilon \in (\varepsilon_0, H)$ the solution $x(t; t_0, x_0)$ may only leave the set B_ζ after crossing all surfaces S_ζ, when ζ takes on values from the interval $[\varepsilon_0, \varepsilon]$.

Let $\eta = b^{-1}(a(\varepsilon_0))$. According to condition (1a) of Theorem 4.4.3, $B_\eta \subset S_{\varepsilon_0} \cup \mathcal{V}_{\varepsilon_0}(t)$. Let the solution $x(t; t_0, x_0)$, beginning in the set B_η at a point of time $t = t_0$, cross the surface S_{ε_0} at a point of time $t^* \geq t_0$. Let t_ζ denote the point of time when the solution $x(t_\zeta; t_0, x_0)$ reaches the surface S_ζ. Consider the behavior of the function $v(t, x)$ along the interval of the path $x(t; t_0, x_0)$ with its origin at the point (t_ζ, x_ζ) and the end at the point $(t_{\zeta+d\zeta}, x_{\zeta+d\zeta})$, located on the surface $S_{\zeta+d\zeta}$. From the expression of the full derivative of the function $v(t, x)$ in view of the system (4.4.11) obtain

$$v(t_{\zeta+d\zeta}, x_{\zeta+d\zeta}) \leq v(t_\zeta, x_\zeta) + \int\limits_{t_\zeta}^{t_{\zeta+d\zeta}} |(\partial v_0/\partial x)^{\mathrm{T}}(f - f^0)| \, dt \int\limits_{t_\zeta}^{t_{\zeta+d\zeta}} |\varphi(y, x)| \, dt.$$

Taking into account the definition of the surface S_ζ, under conditions (2) and (3) of Theorem 4.4.3 obtain

$$a(\zeta + d\zeta) - a(\zeta) = a'(\zeta)d\zeta \leq \lambda(t)w(\zeta)dt + \varkappa(t)c(\zeta)dt.$$

Since $a(\cdot)$ is a strictly monotone increscent function, $a'(\zeta) > 0$ and

$$\frac{d\zeta}{dt} \leq \lambda(t)\frac{d\zeta}{dt}\frac{w(\zeta)}{a'(\zeta)} + \varkappa(t)\frac{c(\zeta)}{a'(\zeta)}. \tag{4.4.14}$$

Keeping the same notation for the variables and passing on to the equation (4.4.13) from the inequality (4.4.14), we see that at all values of t for which $\zeta(t)$ is positive and satisfies the equation (4.4.13) the solution $x(t; t_0, x_0)$ will remain in the time-varying set $S_\zeta \cup \mathcal{V}_\zeta \subset B_\zeta$.

Theorem 4.4.3 is proved.

Remark 4.4.2 If $f^0(t, x) \equiv f(t, x)$, then the nonlinear equation (4.4.13) takes the form

$$\frac{d\zeta}{dt} = \varkappa(t)\frac{c(\zeta)}{a'(\zeta)}.$$

Hence obtain

$$\int\limits_{\varepsilon_0}^{\varepsilon} \frac{a'(\zeta)d\zeta}{c(\zeta)} = \int\limits_{t_0}^{t} \varkappa(t) \, dt. \tag{4.4.15}$$

If in condition (2) of Theorem 4.4.3 we assume $|\varphi(t, x)| \leq \mu c(\|x\|)$, that is, $\varkappa(t) \equiv 1$, then (4.4.15) implies the estimate T from Theorem 4.3.3.

Corollary 4.4.1 Let the following assumptions be taken into account in conditions (1a), (2), and (3):

$$a\|x\|^{r_1} \leq v(t, x) \leq b\|x\|^{r_1},$$
$$\|g(t, x)\| \leq \varkappa_1(t)\|x\|^{r_2},$$
$$\|f(t, x) - f^0(\beta, x)\| \leq \lambda_1(t)\|x\|^{r_3}.$$

Then the equation (4.4.13) takes the form

$$\frac{d\zeta}{dt} = \lambda(t)\zeta^{r_3} + \varkappa(t)\zeta^{r_2}, \tag{4.4.16}$$

$$\zeta(t_0) = \left(\frac{b}{a}\right)^{\frac{1}{r_1}} \|x_0\|, \tag{4.4.17}$$

where the functions $\lambda(t)$ and $\varkappa(t)$ are determined from the inequalities

$$|(\partial v/\partial x)^{\mathrm{T}}(f - f^0)| \leq ar_1\lambda(t)\|x\|^{r_1 + r_3 - 1},$$
$$|\varphi(t, x)| \leq ar_1\varkappa(t)\|x\|^{r_1 + r_2 - 1},$$

and the solution $x(t; t_0, x_0)$ remains in the set B_ζ at all values of t, for which the Cauchy problem (4.4.16) and (4.4.17) has a positive solution $\zeta(t; t_0, \zeta_0)$.

The proof is made by direct substitution of the considered functions into the equation (4.4.13) taking into account that $a(\zeta) = a \cdot \zeta^{r_1}$.

Now consider some particular cases of the problem (4.4.16) and (4.4.17).

Case A. The right-hand parts of the limiting system (4.4.12) and the shortened system coincide. Here $\lambda(t) \equiv 0$ and the equation (4.4.16) takes the form

$$\frac{d\zeta}{dt} = \varkappa(t)\zeta^{r_2}, \tag{4.4.18}$$

the value $\zeta(t_0)$ is determined by the expression (4.4.17). Hence at $r_2 = 1$ obtain

$$\zeta(t) = \left(\frac{b}{a}\right)^{\frac{1}{r_1}} \|x_0\| \exp\left(\int_{t_0}^{t} \varkappa(s)\,ds\right),$$

and at $r_2 \neq 1$

$$\zeta(t) = \left[\left(\frac{b}{a}\right)^{\frac{q}{r_1}} \|x_0\|^q + q \int_{t_0}^{t} \varkappa(s)\,ds\right]^{\frac{1}{q}}, \quad q = 1 - r_2.$$

The value of τ in the estimate $[t_0, t_0 + \tau)$ is determined by the expression

$$\tau = \sup\{t \in J: \zeta(t) \in (0, H)\}. \tag{4.4.19}$$

Remark 4.4.3 The result similar to the obtained one holds if in the system (4.4.11) $g(t,x) \equiv 0$ or $r_2 = r_3$.

Case B. The shortened system is linear, that is, $f(t,x) = A(t)x$, where $A(t)$ is an $(n \times n)$-matrix with its elements continuous and bounded on J. Here $f^0(t,x) = A^0(t)x$, $r_3 = 1$, and the equation (4.4.16) takes the form

$$\frac{d\zeta}{dt} = \lambda(t)\zeta + \varkappa(t)\zeta^{r_2}. \tag{4.4.20}$$

By the substitution $\zeta = \eta^{\frac{1}{1-r_2}}$ the equation (4.4.20) is reducible to a linear equation and integrable by quadratures. Obtain

$$\zeta(t) = E(t)\left\{\left[\left(\frac{b}{a}\right)^{\frac{1}{2}}\|x_0\|\right]^q + q\int_{t_0}^t \varkappa(s)E(-qs)\,ds\right\}^{\frac{1}{q}}, \tag{4.4.21}$$

where

$$E(t) = \exp\left[\int_{t_0}^t \lambda(s)\,ds\right], \quad q = 1 - r_2.$$

The value of τ in the estimate of the interval on which $\|x(t)\| \leq \zeta(t)$, that is, $x(t; t_0, x_0) \in B_\zeta$, is estimated by the formula (4.4.19), taking into account (4.4.21).

Remark 4.4.4 Since the shortened system is linear, the Lyapunov function $v(t,x)$ is chosen in the quadratic form and therefore $r_2 = 2$.

Example 4.4.2 Consider the system

$$\frac{dx_1}{dt} = \alpha\left(\frac{\beta}{t} - 1\right)x_1 + \omega x_2,$$
$$\frac{dx_2}{dt} = -\omega x_1 + \mu\omega\cos^2\nu t \cdot x_1, \tag{4.4.22}$$

where μ is a small parameter, α, β, ω, ν are positive constants, and $t > 0$. For the shortened system (let $\mu = 0$)

$$\frac{dx_1}{dt} = \alpha\left(\frac{\beta}{t} - 1\right)x_1 + \omega x_2,$$
$$\frac{dx_2}{dt} = -\omega x_1 \tag{4.4.23}$$

the full derivative of the Lyapunov function $v = x_1^2 + x_2^2$ is nonnegative; therefore, Theorem 4.3.3 cannot be applied to the system (4.4.22). Taking into account that

$$A^0 = \lim_{t\to\beta}\begin{pmatrix} \alpha\left(\frac{\beta}{t} - 1\right) & \omega \\ -\omega & 0 \end{pmatrix} = \begin{pmatrix} 0 & \omega \\ -\omega & 0 \end{pmatrix}$$

and the state $x_1 = x_2 = 0$ of the respective limiting system is stable (not asymptotically), Theorem 4.4.3 is applicable to it.

Choosing the norm of vector $x \in R^2$ in the form $\|x\| = \max\{|x_1|, |x_2|\}$, for the constants obtain the values $r_1 = 2$, $r_2 = r_3 = 1$, $a = 1$, $b = 2$. The functions $\varkappa(t)$ and $\lambda(t)$ have the form

$$\varkappa(t) = \frac{1}{2}\mu\omega(1 + \cos 2t), \quad \lambda(t) = \alpha\left(\frac{\beta}{t} - 1\right).$$

Applying Theorem 4.4.3 in Case A, obtain $\|x(t)\| \le \zeta(t)$ at all $t \in [t_0, t_0 + \tau)$, $t_0 > 0$, where

$$\zeta(t) = \sqrt{2}\|x_0\| \left(\frac{t}{t_0}\right)^{\alpha\beta} \exp\left[\left(\frac{1}{2}\mu\omega - \alpha\right)(t - t_0) + \frac{1}{4}\mu\omega(\sin 2t - \sin 2t_0)\right]$$

and τ is determined by the formula (4.4.19).

Remark 4.4.5 The interval of the stay of motion on a time-varying set is determined by the inverse transformation of the function $\zeta(t)$.

4.5 Systems with Nonasymptotically Stable Subsystems

Consider a system of the form

$$\frac{dx_s}{dt} = f_s(t, x_s) + \mu g_s(t, x_1, \dots, x_m), \quad s = 1, \dots, m, \tag{4.5.1}$$

where $(x_1^T, \dots, x_m^T)^T = x$, $x_s \in R^{n_s}$, $\sum_{s=1}^{m} n_s = n$, $f_s(t, 0) = g_s(t, 0, \dots, 0) = 0$.
A peculiar property of the system (4.5.1) is that at $\mu = 0$ it falls into m independent subsystems

$$\frac{dx_s}{dt} = f_s(t, x_s), \quad s = 1, \dots, m. \tag{4.5.2}$$

The vector functions $f_s(t, x_s)$ and $g_s(t, x)$, $s = 1, \dots, m$, in the domain Ω satisfy the condition of the existence and uniqueness of the solution of the Cauchy problem for the systems (4.5.1) and (4.5.2), in addition, $f_s(t, x_s)$, $s = 1, \dots, m$, satisfy the Lipschitz condition with respect to x_s, $s = 1, \dots, m$, with the constant L.

Assume that the solution $x_s = 0$ of the systems (4.5.2) is stable uniformly with respect to t_0 (nonasymptotically) and the general solution is known $\overline{x}_s(t) = \overline{x}_s(t, t_0, x_{s0})$, $s = 1, \dots, m$, $(t_0, x_0) \in \text{int } \Omega$.

Assume that for each subsystem (4.5.2) a continuously differentiable Lyapunov function $v_s(t, x_s)$ is known, which has the respective properties.

Let the vectors x_s, $s = 1, \ldots, m$, be numbered in an order convenient for the study, let α_s, $s = 1, \ldots, m$, be some sequence for which $\alpha_1 = 1$, and let each subsequent term α_k, $k = 2, \ldots, m$, be equal either to the previous term or to the index k. Denote

$$\varphi_s^{\alpha_1}(t, x_1, \ldots, x_m) = \left(\frac{\partial v_s}{\partial x_s}\right)^{\mathrm{T}} g_s(t, x_1, \ldots, x_m),$$

$$\varphi_s^{\alpha_s}(t, x_{\alpha_s}, \ldots, x_m) = \left(\frac{\partial v_s}{\partial x_s}\right)^{\mathrm{T}} g_s(t, 0, \ldots, 0, x_{\alpha_s}, \ldots, x_m),\qquad(4.5.3)$$

$$s = 1, \ldots, m.$$

Consider the sequence of means

$$\Theta_s^{\alpha_s}(t_0, x_{\alpha_s 0}, \ldots, x_{m0}) = \lim_{T \to \infty} \frac{1}{T} \int_{t_0}^{t_0+T} \varphi_s^{\alpha_s}(t, \overline{x}_{\alpha_s}(t), \ldots, \overline{x}_m(t)) dt,\qquad(4.5.4)$$

$$s = 1, \ldots, m.$$

Let $v_s^*(x_s)$ denote nonpositive functions and let $w_s^*(x_s)$ denote nonnegative functions defined and continuous in the domains $D_s = \{x_s\colon \|x_s\| < H\}$, $s = 1, 2, \ldots, m$.

The mean $\overline{\varphi}_s^{\alpha_s}$ is less than zero in the set $E(v_s^* = 0)$, if for any numbers η_s and ε_s, $0 < \eta_s < \varepsilon_s < H$, one can find positive quantities $r_s(\eta_s, \varepsilon_s)$ and $\delta_s(\eta_s, \varepsilon_s)$ such that $\Theta_s^{\alpha_s}(t_0, x_{\alpha_s 0}, \ldots, x_{m0}) < -\delta_s(\eta_s, \varepsilon_s)$ at $\eta_s \leq \|x_{s0}\| \leq \varepsilon_s$, $\rho(x_{s0}, E(v_s^* = 0)) < r_s(\eta_s, \varepsilon_s)$ for all $t_0 \in R_+$, $x_{k0} \in D_k$, $k = \alpha_s, \ldots, s - 1, s + 1, \ldots, m$.

Consider the following statement.

Theorem 4.5.1 *Let the following conditions be satisfied for the system (4.5.1):*

(1) *there exist positive definite decrescent functions $v_s(t, x_s)$, $s = 1, \ldots, m$, and the following inequalities hold:*

$$\frac{\partial v_s}{\partial t} + \left(\frac{\partial v_s}{\partial x_s}\right)^{\mathrm{T}} f_s(t, x_s) \leq v_s^*(x_s) \leq 0, \quad s = 1, \ldots, m;$$

(2) *there exist summable functions $M_s(t)$ and $F_s(t)$, constants M_{s0}, F_{s0}, and functions $\chi_s(\beta) \in K$ such that on any finite interval the following inequalities will hold:*

$$|\varphi_s^{\alpha_1}(t, x_1', \ldots, x_m') - \varphi_s^{\alpha_1}(t, x_1'', \ldots, x_m'')| \leq F_s(t)\chi_s\left(\sum_{k=1}^{m} |x_k' - x_k''|\right),$$

$$\int\limits_{t_1}^{t_2} F_s(t)dt \leq F_{s0}(t_2 - t_1),$$

$$\|g_s(t, x_1, \ldots, x_m) \leq M_s(t), \quad \int\limits_{t_1}^{t_2} M_s(t)dt \leq M_{s0}(t_2 - t_1),$$

$$s = 1, \ldots, m;$$

(3) *uniformly with respect to* $t_0, x_{\alpha_s 0}, \ldots, x_{m0}$ *there exist means*

$$\Theta_s^{\alpha_s}(t_0, x_{\alpha_s 0}, \ldots, x_{m0}), \quad s = 1, \ldots, m;$$

(4) *the mean* $\Theta_s^{\alpha_s}$ *is less than zero in the set* $E(v_s^* = 0)$, $s = 1, \ldots, m$.

Then the solution $x = 0$ *of the system (4.5.1) is* μ-*stable.*

Proof Let $\varepsilon \in (0, H)$ and $t_0 \in R_+$ be specified. We will show that for ε it is possible to find positive numbers $\eta(\varepsilon)$ and $\mu_0(\varepsilon)$, not depending on t_0, such that any solution $x_s(t, t_0, x_0)$, $s = 1, \ldots, m$, of the system (4.5.1) will satisfy the condition

$$\sum_{s=1}^{m} \|x_s(t, t_0, x_0)\| < \varepsilon$$

for all $t > t_0$ at $\sum\limits_{s=1}^{m} \|x_{s0}\| < \eta(\varepsilon)$ and $\mu \in (0, \mu_0)$.

Choose $\varepsilon_m = \varepsilon/m$. In view of condition (1) of the theorem, for the function $v_m(t, x_m)$ there exist functions $a_m(\|x_m\|) \in K$ and $b_m(\|x_m\|) \in K$ such that

$$a_m(\|x_m\|) \leq v_m(t, x_m) \leq b_m(\|x_m\|). \tag{4.5.5}$$

Let $\eta_m(\varepsilon_m) = b_m^{-1}(a_m(\varepsilon_m/2))$. Then all points of the moving surface $v_m(t, x_m) = a(\varepsilon_m/2)$ for all $t \in R_+$ will satisfy the inequality $\eta_m \leq \|x_m\| \leq \varepsilon_m/2$. For the numbers η_m and $\varepsilon_m/2$ according to condition (4), there exists a value $\delta_m(\eta_m, \varepsilon_m/2)$.

Determine the number

$$\varepsilon_{m-1} = \min\left(\frac{\chi_m^{-1}(\delta_m/4F_{m0})}{2(\alpha_m - 1)}, \ \varepsilon_m\right).$$

For $v_{m-1}(t, x_{m-1})$ there exist functions $a_{m-1}(\|x_{m-1}\|)$ and $b_{m-1}(\|x_{m-1}\|)$ from the K-class such that a relation similar to (4.5.5) will hold. Determine the number

$$\eta_{m-1}(\varepsilon_{m-1}) = b_{m-1}^{-1}(a_{m-1}(\varepsilon_{m-1}/2)).$$

Under condition (4) for η_{m-1} and $\varepsilon_{m-1}/2$ it is possible to determine the quantity $\delta_{m-1}(\eta_{m-1}, \varepsilon_{m-1}/2)$. Assume that

$$\varepsilon_{m-2} = \min\left(\frac{\chi_{m-1}^{-1}(\delta_{m-1}/4F_{m-10})}{2(\alpha_{m-1} - 1)}, \ \varepsilon_{m-1}\right).$$

Continuing this process, for each function $v_s(t, x_s)$, $s = m - 2, \ldots, 1$, determine the respective functions $a_s(\|x_s\|)$, $b_s(\|x_s\|)$ and numbers

$$\varepsilon_s = \min \left(\frac{\chi_{s+1}^{-1}(\delta_{s+1}/4F_{s+10})}{2(\alpha_{s+1} - 1)}, \ \varepsilon_{s+1} \right),$$

$$\eta_s(\varepsilon_s) = b_s^{-1}(a_s(\varepsilon_s/2)), \quad \delta_s(\eta_s, \varepsilon_s/2).$$

Considering the components $x_s(t, t_0, x_0)$ in increscent order of the index s, $s = 1, \ldots, m$, show that each of them satisfies the condition $\|x_s(t)\| < \varepsilon_s$ at $\mu < \mu_s$ for all $t \geq t_0$ at $\|x_{k0}\| < \eta_k$, $k = 1, \ldots, m$. The proofs of this statement for each component are similar; therefore, we will give proof for $x_s(t)$, assuming that for $x_1(t), \ldots, x_{s-1}(t)$ it has been proved already and the components $x_i(t)$, $i = s + 1, \ldots, m$, at $t > t_0$ remain in the domain D_i, $i = s + 1, \ldots, m$. Assume that the component $x_s(t, t_0, x_0)$ left the domain $\|x_s\| < \eta_s$ and at a point of time $t = t'_{s0}$ crossed the surface $v_s(t, x_s) = a_s(\varepsilon_s/2)$ in a point x'_{s0}. For this point, in view of condition (4), one of the following inequalities will hold:

$$\rho(x'_{s0}, E(v_s^* = 0)) \leq r_s(\eta_s, \varepsilon_s/2),$$

$$\Theta_s^{\alpha_s}(t'_0, x'_{\alpha_s 0}, \ldots, x'_{m0}) < -\delta_s(\eta_s, \varepsilon_s/2).$$

Here $x'_{k0} = x_k(t'_0, t_0, x_0)$, $k = \alpha_s, \ldots, m$.

Consider some properties of the solutions $x_s(t, t_0, x_0)$.

(a) Let the following conditions be satisfied at a point of time τ

$$v_s(\tau, x_s(\tau)) = a_s(\varepsilon_s/2),$$

$$\rho(x_s(\tau), E(v_s^* = 0)) \geq r_s(\eta_s, \varepsilon_s/2).$$

Taking into account the equality $g_s(t, 0, \ldots 0) = 0$, from condition (2) of the theorem obtain

$$|\varphi_s^{\alpha_1}(t, x_1, \ldots, x_m)| \leq c_s F(t)$$

at $(t, x) \in \Omega$, where c_s is some constant. Choosing

$$\mu < \mu'_s = \gamma/2c_s F_{s0}, \quad \gamma = \inf_{x_s \in P_s} |v_s^*(x_s)|,$$

$$P_s = \{x \colon \rho(x_s, E(v_s^* = 0)) \geq r_s/2, \ \eta_s \leq \|x_s\| \leq \varepsilon_s\},$$

for all $t > \tau$, at which the conditions

$$\|x_s(t)\| \geq \eta_s, \quad \rho(x_s(t), E(v_s^* = 0)) \geq r_s/2$$

are satisfied, obtain

$$v_s(t, x_s(t)) \leq v_s(\tau, x_s(\tau)) + \int_\tau^t v_s^*(x_s(t)) \, dt + \mu \int_\tau^t \varphi_s^{\alpha_1}(t, x_1(t), \ldots, x_m(t)) \, dt$$

$$\leq v_s(\tau, x_s(\tau)) - \frac{\gamma_s}{2}(t - \tau).$$

$$(4.5.6)$$

The function $v_s(t, x_s(t))$ is not increscent and, respectively, $x_s(t, \tau, x(\tau))$ will not leave the domain bounded by the surface $v_s(t, x_s) = a_s(\varepsilon_s/2)$, which means that for it the inequality will not be violated at least as long as the condition

$$\rho(x_s(t), E(v^* = 0)) \geq r_s/2$$

holds.

(b) Let the following conditions be satisfied at a point of time τ:

$$v_s(\tau, x_s(\tau)) = a_s(\varepsilon_s/2),$$
$$\rho(x_s(\tau), E(v_s^* = 0)) < r_s(\eta_s, \varepsilon_s/2).$$

In view of condition (4) of Theorem 4.5.1

$$\Theta_s^{\alpha_s}(\tau, x_{\alpha_s}(\tau), \ldots, x_m(\tau)) < -\delta_s(\eta_s, \varepsilon_s/2). \qquad (4.5.7)$$

Taking this into account, estimate the change of the function $v_s(t, x_s)$. Represent the last integral in the inequality (4.5.6) in the form

$$\int_\tau^t \varphi_s^{\alpha_1}(t, x_1(t), \ldots, x_m(t)) \, dt = \int_\tau^t [\varphi_s^{\alpha_1}(t, x_1(t), \ldots, x_m(t))$$

$$\qquad (4.5.8)$$

$$- \varphi_s^{\alpha_s}(t, \overline{x}_{\alpha_s}(t), \ldots, \overline{x}_m(t))] \, dt + \int_\tau^t \varphi_s^{\alpha_s}(t, \overline{x}_{\alpha_s}(t), \ldots, x_m(t)) \, dt,$$

where $\overline{x}_k(t) = \overline{x}_k(t, \tau, x_k(\tau))$, $k = \alpha_s, \ldots, m$. Estimate the last integral in the expression (4.5.8), using the estimate of the mean $\Theta_s^{\alpha_s}(\tau, x_{\alpha_s}(\tau), \ldots, x_m(\tau))$ in the inequality (4.5.7) and choosing the time interval l_s so large that at $t > \tau + l_s$ the following condition will be satisfied:

$$\int_\tau^t \varphi_s^{\alpha_s}(t, \overline{x}_{\alpha_s}(t), \ldots, \overline{x}_m(t)) \, dt < -\frac{3}{4} \delta_s(t - \tau). \qquad (4.5.9)$$

Choose μ_s'' so small that on the interval $[\tau, \tau + 2l_s]$ at $\mu < \mu_s''$ the following inequalities will hold:

$$\|x_s(t) - \overline{x}_s(t)\| < \frac{\varepsilon_s}{2}, \quad \sum_{k=\alpha_s}^m \|x_k - \overline{x}_k\| < \frac{\chi_s^{-1}(\delta_s/4F_{s0})}{2}. \qquad (4.5.10)$$

Using Lemma 4.2.2, determine μ_s'':

$$\mu_s'' = \min \left(\frac{\varepsilon_s}{4M_{s0}l_s \exp(2l_s L)}, \chi_s^{-1}(\delta_s/4F_{s0})/4l_s \exp(2Ll_s) \sum_{k=\alpha_s}^m M_{s0} \right).$$

The first inequality of (4.5.10) means that $x_s(t, \mu)$ at $\mu < \mu_s''$ will not leave the domain $\|x_s\| < \varepsilon$ at least on the time interval $2l_s$.

Now estimate the second integral in the expression (4.5.8), using condition (2) of the theorem, the second inequality of (4.5.10), and the choice of numbers ε_i, $i = 1, \ldots, \alpha_s - 1$. For $t \in [\tau, \tau + 2l_s]$ and $\mu < \mu_s'' = \min(\mu_s'', \mu_{\alpha_s - 1})$ obtain

$$\int_\tau^t |\varphi_s^{\alpha_1}(t, x(t)) - \varphi_s^{\alpha_s}(t, \bar{x}_{\alpha_s}(t), \ldots, \bar{x}_m(t))\, dt$$

$$\leq \int_\tau^t F_s(t)\chi_s \left(\sum_{i=1}^{\alpha_s - 1} \|x_i(t)\| + \sum_{k=\alpha_s}^m \|x_k(t) - \bar{x}_k(t)\| \right) dt \qquad (4.5.11)$$

$$\leq \int_\tau^t F_s(t)\chi_s(\chi_s^{-1}(\delta_s/4F_{s0}))\, dt \leq \delta_s(t - \tau)/4.$$

Substituting the estimates (4.5.9) and (4.5.11) into (4.5.8), for $t \in [\tau + l_s, \tau + 2l_s]$ and $\mu < \mu_s''$ obtain

$$\int_\tau^t \varphi_s^{\alpha_1}(t, x(t))\, dt < -\frac{\delta_s}{2}(t - \tau).$$

Thus, in the inequality (4.5.6) the last integral becomes negative at least from the point of time $t = \tau + l_s$. Therefore, the component $x_s(t, \mu)$, having left the domain bounded by the surface $v_s(t, x_s) = a_s(\varepsilon_s/2)$, in view of the chosen μ, will remain in the domain $\|x_s\| < \varepsilon_s$ and at some point of time from the interval $(\tau, \tau + 2l_s)$ will return into the domain bounded by the surface $v_s(t, x_s) = a_s(\varepsilon_s/2)$.

Choose $\mu_s = \min(\mu_s', \mu_s'')$. Then at $\mu < \mu_s$ properties (a) and (b) imply that for all $t \geq t_0$ the component $x_s(t, t_0, x_0)$ will remain in the domain $\|x_s\| < \varepsilon_s$.

Thus, considering the components $x_s(t, \mu)$ of the solution in increscent order of the index s, $s = 1, \ldots, m$, we see that each of them satisfies the condition $\|x_s(t)\| < \varepsilon_s$ at $\|x_{k0}\| < \eta_k$, $k = 1, \ldots, m$, and $t \geq t_0$. Choosing $\mu_0 = \min(\mu_s)$, $\eta(\varepsilon) = \min(\eta_s)$, $s = 1, \ldots, m$, and taking into account the chosen numbers ε_s, $s = 1, \ldots, m$, obtain

$$\sum_{s=1}^m \|x_s(t, \mu)\| < \varepsilon \quad \text{at all} \quad t \geq t_0 \quad \text{as soon as,} \quad \sum_{s=1}^m \|x_{s0}\| < \eta.$$

Here the quantities $\eta(\varepsilon)$ and $\mu_0(\varepsilon)$ were chosen irrespective of t_0.

Theorem 4.5.1 is proved.

If we consider the functions $g_s(t, x_1, \ldots, x_m)$, $s = 1, \ldots, m$, as weak connections such that $g_s(t, 0, \ldots, 0) = 0$, $s = 1, \ldots, m$, then for the correctness of

Theorem 4.5.1 one should additionally require the correctness of the following
inequalities on any finite interval $[t_1, t_2]$

$$\varphi_s^{\alpha_1}(t, x_1, \ldots, x_m) \le N_s(t), \quad \int\limits_{t_1}^{t_2} N_s(t)\, dt \le N_{s0}(t_2 - t_1)$$

at $t \in R_+$, $x_s \in D_s \setminus E(v_s^* = 0)$, $x_k \in D_k$, N_{s0} is a constant, $s = 1, \ldots, m$,
$k = 1, \ldots, s-1, s+1, \ldots, m$.

Here are some applications of Theorem 4.5.1.

Example 4.5.1 Consider the system

$$
\begin{aligned}
\frac{dx_1}{dt} &= -x_1 + x_2^2 + \mu x_2 y_1, & x_1(t_0) &= x_{10}, \\
\frac{dx_2}{dt} &= \mu(x_2 y_2 - x_1 x_2 + x_1 z + x_1^2 \sin t), & x_2(t_0) &= x_{20}, \\
\frac{dy_1}{dt} &= -y_1 + y_2 + \mu x_2^2, & y_1(t_0) &= y_{10}, \qquad (4.5.12) \\
\frac{dy_2}{dt} &= \mu[-x_2^2 y_2 z - y_1 y_2 + (y_1 + y_2) \cos t], & y_2(t_0) &= y_{20}, \\
\frac{dz}{dt} &= \mu(-z^3 + y_1 y_2 - x_1^2 z + z^2 \cos t), & z(t_0) &= z_0,
\end{aligned}
$$

which falls into three subsystems at $\mu = 0$:

$$
\begin{aligned}
\frac{dx_1}{dt} &= -x_1 + x_2^2, & \frac{dx_2}{dt} &= 0, \\
\frac{dy_1}{dt} &= -y_1 + y_2, & \frac{dy_2}{dt} &= 0, \qquad (4.5.13) \\
\frac{dz}{dt} &= 0. &
\end{aligned}
$$

Write the solution of the subsystems (4.5.13) in the form

$$
\begin{aligned}
\overline{x}_1(t) &= x_{20}^2 + (x_{10} - x_{20}^2)\exp[-(t - t_0)], & \overline{x}_2(t) &= x_{20}, \\
\overline{y}_1(t) &= y_{20} + (y_{10} - y_{20})\exp[-(t - t_0)], & \overline{y}_2(t) &= y_{20}, \\
\overline{z}(t) &= z_0.
\end{aligned}
$$

Start the investigation of the μ-stability of the system (4.5.12) from the
second subsystem. Choose the Lyapunov function $v_1(y) = y_2^2 + (y_1 - y_2)^2$,
whose derivative in view of the second subsystem of (4.5.13) satisfies the relation $\dot{v}_1(y) = -2(y_1 - y_2)^2 = v_1^*(y) \le 0$. Calculating the mean for that
subsystem, obtain $\Theta_1^1(y_0, z_0, x_0) = -2y_{20}^4 - 2y_{20}^2 x_{20}^2 z_0^2$. Obviously, the mean
Θ_1^1 is less than zero in the set $E(v_1^* = 0) = \{y\colon y_1 = y_2\}$ at any x_{20} and z_0.

For the third subsystem take the Lyapunov function $v_2(z) = z^2$. Its derivative in view of the third subsystem of (4.5.13) is zero. Assuming $y_1 = 0$ and

$y_2 = 0$, calculate the mean for that subsystem $\Theta_2^2(z_0, x_0) = -2(z_0^4 + z_0^2 x_{20}^4)$. The mean $\overline{\varphi}_2^2$ is negative definite with respect to z_0 at any x_{20} and therefore it satisfies condition (4) of Theorem 4.5.1.

Now for the first subsystem choose the function $v_3(x) = x_2^2 + (x_1 - x_2)^2$, whose full derivative satisfies the relation $\dot{v}_3(x) = -2(x_1 - x_2^2)^2 = v_3^*(x) \leq 0$. At $y_1 = 0$, $y_2 = 0$, $z = 0$ obtain $\Theta_3^3(x_0) = -2x_{20}^4$.

The conditions of Theorem 4.5.1 are satisfied; therefore, the solution $y = 0$, $z = 0$, $x = 0$ of the sytem (4.5.12) is μ-stable.

Now we will need the following definition.

The mean $\Theta_s^{\alpha_s}(t_0, x_{\alpha_s 0}, \ldots, x_{m0}) > 0$ in the set $E(w_s^* = 0)$ of the domain $v_s > 0$, if for any positive λ, however small it may have been chosen, there exist positive numbers $\delta_s(\lambda)$ and $r_s(\lambda)$ such that at $t_0 \in R_+$ and $x_{s0} \in D_s$ satisfying the condition $v_s(t_0, x_0) > \lambda$, $\rho(x_{s0}, E(w^* = 0)) < r_s(\lambda)$, the inequality

$$\Theta_s^{\alpha_s}(t_0, x_{\alpha_s 0}, \ldots, x_{m0} > \delta_s(\lambda)$$

will hold at all $x_{k0} \in D_k$, $k = \alpha_s, \ldots, s-1, s+1, \ldots, m$.

Theorem 4.5.2 *Let the following conditions be satisfied for the system (4.5.1):*

(1) *for the s-th subsystem there exists a function $v_s(t, x_s)$ which has a domain $v_s(t, x_s) > 0$ and is bounded therein;*

(2) *in the domain $v_s > 0$*

$$\frac{\partial v_s}{\partial t} + \left(\frac{\partial v_s}{\partial x_s}\right)^{\mathrm{T}} f_s(t, x_s) \geq w_s^* \geq 0;$$

(3) *there exist summable functions $M_s(t)$, $F_s(t)$, constants M_{s0}, F_{s0}, and a function $\chi_s(\beta) \in K$ such that on any finite interval $[t_1, t_2]$ the following inequalities will hold:*

$$\|g_s(t, x_1, \ldots, x_m)\| \leq M_s(t), \quad \int_{t_1}^{t_2} M_s(t)\, dt \leq M_{s0}(t_2 - t_1),$$

$$|\varphi_s^{\alpha_1}(t, x_1', \ldots, x_m') - \varphi_s^{\alpha_1}(t, x_1'', \ldots, x_m'')| \leq F_s(t)\chi_s\left(\sum_{k=1}^{m} \|x_k' - x_k''\|\right),$$

$$\int_{t_1}^{t_2} F_s(t)\, dt \leq F_{s0}(t_2 - t_1)$$

at t, x_s from the domain $v_s > 0$ and $x_n \in D_n$, $n = 1, \ldots, s-1, s+1, \ldots, m$;

(4) *uniformly with respect to* (t_0, x_{s0}) *from the domain* $v_s > 0$ *and* $x_{n0} \in D_n$, $n = \alpha_s, \ldots, s-1, s+1, \ldots, m$, *there exists a mean* $\Theta_s^{\alpha_s}(t_0, x_{\alpha_s 0}, \ldots, x_{m0}$;

(5) *the mean* $\Theta_s^{\alpha_s}$ *is above zero in the set* $E(v_s^* = 0)$ *of the domain* $v_s > 0$;

(6) *for each i-th subsystem* $(i = 1, \ldots, s-1)$ *the conditions of Theorem 4.5.1 are satisfied.*

Then the solution $x = 0$ *of the system (4.5.1) is* μ-*unstable.*

Proof Let $\varepsilon_s \in (0, H)$ and $t_0 \in R_+$ be specified. We will show that it is possible to find an arbitrarily small x_0 such that the solution $x(t, t_0, x_0)$ of the system (4.5.1) at some point of time will leave the domain $\|x\| = \sum_{k=1}^{m} \|x_k\| < \varepsilon$. Choose the value of x_{s0} as small as we please and such that at the specified t_0 that value would belong to the domain $v_s > 0$. Then there will exist a number $\alpha > 0$ such that $v_s(t_0, x_{s0}) > \alpha$. Under condition (5) of the theorem, for $\alpha > 0$ there exist positive numbers $\delta_s(\alpha)$ and $r_s(\alpha)$ such that one of the following inequalities holds:

$$\rho(x_{s0}, E(v_s^* = 0)) > r_s(\alpha)$$

or

$$\Theta_s^{\alpha_s}(t_0, x_{\alpha_s 0}, \ldots, x_{m0} > \delta_s(\alpha).$$

Determine the number

$$\varepsilon_{s-1} \leq \min \left(\frac{\chi_s^{-1}(\delta_s/4F_{s0})}{2(\alpha_s - 1)}, \ \varepsilon_s \right).$$

For the positive definite function $v_{s-1}(t, x_{s-1})$ there exist functions $a_{s-1}(\|x_{s-1}\|)$ and $b_{s-1}(\|x_{s-1}\|)$ from the class K such that the following inequality would hold:

$$a_{s-1}(\|x_{s-1}\|) \leq v_{s-1}(t, x_{s-1}) \leq b_{s-1}(\|x_{s-1}\|).$$

Choose $\eta_{s-1} = b_{s-1}^{-1}(a_{s-1}(\varepsilon_{s-1}/2))$. For the numbers η_{s-1} and $\varepsilon_{s-1}/2$, according to condition (4), there exists a value $\delta_{s-1}(\eta_{s-1}, \varepsilon_{s-1}/2)$. Thus, for each subsystem sequentially determine

$$\varepsilon_k = \min \left(\frac{\chi_{k+1}^{-1}(\delta_{k+1}/4F_{k+10})}{2(\alpha_{k+1} - 1)}, \ \varepsilon_{k+1} \right),$$

the functions $a_k(\|x_k\|) \in K$ and $b_k(\|x_k\|) \in K$, the numbers $\eta_k = b_k^{-1}(a_k(\varepsilon_k/2))$ and $\delta_k(\eta_k, \varepsilon_k/2)$, $k = s-2, \ldots, 1$.

In the proof of Theorem 4.5.1 it was shown that for $\|x_{i0}\| < \eta_i$, x_{j0} are any arbitrarily small, $i = 1, \ldots, s-1$, $j = s+1, \ldots, m$, each component $x_i(t, t_0, x_{10}, \ldots, x_{m0})$ will satisfy the condition $\|x_i\| < \varepsilon_i$ at $\mu < \mu_0'$, at least until any of the components $x_j(t, t_0, x_0)$, $j = s, s+1, \ldots, m$, leaves the domain $\|x_j\| < H$, $j = s, \ldots, m$. Show that for $x_s(t, t_0, x_0)$ at some point of time the condition $\|x_s\| = \varepsilon_s$ will be satisfied.

Consider some properties of the solution $x_s(t, t_0, x_0)$.

(a) Let the following conditions be satisfied at a point of time τ

$$v_s(\tau, x_s(\tau)) > \alpha, \quad \rho(x_s(\tau), E(w_s^* = 0)) \geq r_s(\alpha).$$

Estimate the change of the function $v_s(t, x_s(t))$. From condition (3) of the theorem obtain $|\varphi_s^{\alpha_1}(t, x_1, \ldots, x_m)| \leq c_s F_s(t)$ at $(t, x) \in \Omega$, where c_s is some constant. Choose

$$\mu < \mu_s' = \gamma_s/2c_s F_{s0}, \quad \gamma_s = \inf_{x_s \in P_s} v(x_s),$$

$$P_s = \{x_s : \rho(x_s, E(w_s^* = 0)) \geq r_s/2,$$

$$v_s(t, x_s) > \alpha, \quad t \in R_+, \quad \|x_s\| < \varepsilon_s\}.$$

Then for all $t > \tau$, at which the conditions

$$x_s \in U = \{x_s : \rho(x_s, E(w_s^* = 0)) \geq r_s/2, \|x_s\| < \varepsilon_s\}$$

are satisfied, obtain

$$v_s(t, x_s(t)) \geq w_s^*(x_s(t)) + \mu\varphi_s^{\alpha_1}(t, x_1(t), \ldots, x_m(t)) \geq \frac{\gamma_s}{2}. \qquad (4.5.14)$$

The function $v_s(t, x_s(t))$ is not decrescent; therefore, the solution $x_s(t, \tau, x(\tau))$ will not leave the domain $v_s > 0$ and the inequality $v_s > \alpha$ will not be violated. Taking into account that in the domain $v_s > 0$ at $\|x_s\| < \varepsilon_s$ in view of condition (1) of the theorem $|(v_s(t, x_s)| < \omega$, where ω is some constant, show that $x_s(t, \mu)$ cannot permanently remain in the domain U within the time interval $T = 2(\omega - \alpha)/\gamma_s$. Indeed, if we assume the contrary, then from the inequality (4.5.14) at a point τ we will obtain

$$v_s(t, x_s(t)) \geq v_s(\tau, x(\tau)) + \frac{\gamma_s}{2}(t - \tau) > \omega.$$

This inequality contradicts the condition of the boundedness of the function $v_s(t, x_s)$.

The contradiction proves that there exists a point of time from the interval $\tau + T$, when one of the inequalities is violated:

$$\rho(x_s, E(w_s^* = 0)) \geq \frac{r_s}{2}, \quad \text{or} \quad \|x_s\| < \varepsilon_s.$$

The violation of the second inequality means the instability. Assume that the first inequality is violated.

(b) Let the following conditions be satisfied at a point of time τ:

$$v_s(\tau, x(\tau)) > \alpha, \quad \rho(x_s(\tau), E(w_s^* = 0)) > r_s(\alpha).$$

Taking into account condition (5) of the theorem, obtain

$$\Theta_s^{\alpha_s}(\tau, x_{\alpha_s}(\tau), \ldots, x_m(\tau)) > \delta_s(\alpha). \qquad (4.5.15)$$

Estimate the change of the function $v_s(t, x_s(t))$. For $t > \tau$ the following condition is satisfied:

$$v_s(t, x_s(t)) \geq v_s(\tau, x(\tau)) + \mu \int_\tau^t \varphi_s^{\alpha_s}(t, x(t))\, dt \geq v_s(\tau, x_s(\tau))$$

$$- \mu \int_\tau^t |\varphi_s^{\alpha_s}(t, x(t)) - \varphi_s^{\alpha_s}(t, \overline{x}_{\alpha_s}(t), \ldots, \overline{x}_m(t))|\, dt \quad (4.5.16)$$

$$+ \mu \int_\tau^t \varphi_s^{\alpha_s}(t, \overline{x}_{\alpha_s}(t), \ldots, \overline{x}_m(t))\, dt.$$

Estimate the last integral in (4.5.16), using the existence of the mean $\overline{\varphi}_s^{\alpha_s}(\tau, x_{\alpha_s}(\tau), \ldots, x_m(\tau))$, the inequality (4.5.15), and choosing the time interval l_s so large that at $t > \tau + l_s$ the following inequality would hold:

$$\int_\tau^t \varphi_s^{\alpha_s}(t, \overline{x}_{\alpha_s}(t), \ldots, \overline{x}_m(t))\, dt > \frac{3}{4}\delta_s(t - \tau). \quad (4.5.17)$$

Choose μ_s'' so that on the interval $[\tau, \tau + 2l_s + T]$ at $\mu < \mu_s''$ the following condition would be satisfied:

$$\sum_{k=\alpha_s}^m \|x_k - \overline{x}_k\| < \frac{\chi_s^{-1}(\delta_s/4F_{s0})}{2}. \quad (4.5.18)$$

This can be done by using Lemma 4.2.2:

$$\mu_s'' = \frac{\chi_s^{-1}(\delta_s/4F_{s0})}{2(2l_s + T)\exp(2l_s + T)\sum\limits_{k=\alpha_s}^m M_{k0}}.$$

From condition (3) of the theorem, taking into account the inequality (4.5.18) and the chosen numbers ε_i, $i = 1, \ldots, \alpha_s - 1$, obtain

$$\int_\tau^t |\varphi_s^{\alpha_1}(t, x(t)) - \varphi_s^{\alpha_s}(t, \overline{x}_{\alpha_s}(t), \ldots, \overline{x}_m(t))\, dt$$

$$\leq \int_\tau^t F_s(t)\chi_s\left(\sum_{i=1}^{\alpha_s-1} \|x_i(t)\| + \sum_{k=\alpha_s}^m \|x_k(t) - \overline{x}_k(t)\|\right) dt \quad (4.5.19)$$

$$\leq \frac{\delta_s}{4}(t - \tau)$$

at $t \in [\tau, \tau + 2l_s + T]$.

Choose μ_s''' so that the solution $x_s(t, \tau, x(\tau))$ on the interval $[\tau, \tau + 2l_s + T]$ would not leave the domain $v_s > 0$. For this purpose, μ_s''' should satisfy the inequality

$$\mu_s'''\left[\int_\tau^t \varphi_s^{\alpha_s}(t, \bar{x}_{\alpha_s}(t), \ldots, \bar{x}_m(t))\, dt + \frac{\delta_s}{4}(t - \tau)\right] < \frac{\alpha}{2}$$

at $t \in [\tau, \tau + 2l_s + T]$. Then at

$$\mu < \mu_0'' = \min(\mu_0', \mu_s'', \mu_s''')$$

and $t \in [\tau + 2l_s, \tau + 2l_s + T]$ from the inequality (4.5.16), taking into account (4.5.17) and (4.5.19), obtain

$$v_s(t, x_s(t)) \geq v_s(\tau, x(\tau)) + \mu\frac{\delta_s}{2}(t - \tau) > \alpha. \qquad (4.5.20)$$

Therefore, due to the chosen μ_0'', the solution $x_s(t, \tau, x(\tau))$ will not leave the domain $v_s > 0$ on the interval $[\tau, \tau + 2l_s + T]$ and at any point $t \in [\tau + 2l_s, \tau + 2l_s + T]$ the estimate (4.5.20) holds.

Choose $\mu_0 = \min(\mu_s', \mu_0'')$ and consider the sequence of points of time

$$t_i = t_0 + i(2l_s + T), \quad i = 1, 2, \ldots.$$

At an initial point of time t_0 the conditions of one of the cases (a) or (b) are satisfied. The function $v_s(t, x_s(t))$ will increase on each interval $[t_i, t_{i+1}]$ at least by the value $\mu\delta_s l_s$ at $\mu \in (0, \mu_0)$. Assuming that $x_s(t, t_0, x_0)$ on an interval $[t_0, t_n]$, where n is the smallest integer satisfying the condition

$$n \geq (\omega - \alpha)/\mu\delta_s l_s,$$

is in the domain $\|x_s\| < \varepsilon_s$, for a point of time t_n obtain

$$v_s(t_n, x_s(t_n)) \geq v_s(t_0, x_{s0}) + n\mu\delta_s l_s.$$

This inequality contradicts the condition of the boundedness of the function $v_s(t, x_s)$ in the domain $v_s > 0$. The contradiction means that there exists a point of time $t_1 \in (t_0, t_0 + n(2l_s + T))$ at which the condition $\|x(t_1, \mu)\| \geq \|x_s(t_1, t_0, x_0)\| = \varepsilon_s$ is satisfied.

The theorem is proved.

If we consider $g_s(t, x_1, \ldots, x_m)$, $s = 1, \ldots, m$, as weak connections for which $g_s(t, 0, \ldots, 0) \neq 0$, then for the correctness of Theorem 4.5.2 additional limitations on $\varphi_s^{\alpha_1}(t, x_1, \ldots, x_m)$ are required, like it was done earlier.

4.6 Stability with Respect to a Part of Variables

Represent the system of equations (4.2.1) in the form

$$\frac{dy}{dt} = Y(t, y, z) + \mu G(t, y, z),$$
$$\frac{dz}{dt} = Z(t, y, z) + \mu Q(t, y, z), \tag{4.6.1}$$

where $y \in R^m$, $z \in R^p$, $Y \in C(R_+ \times R^m \times R^p, R^m)$, $G \in C(R_+ \times R^m \times R^p, R^m)$, $Z \in C(R_+ \times R^m \times R^p, R^p)$, $Q \in C(R_+ \times R^m \times R^p, R^p)$, i.e. $f(t, x) = (Y^T(t, y, z), Z^T(t, y, z))^T$, $x = (y^T, z^T)^T$, $\|y\| = (y^T y)^{1/2}$, $\|z\| = (z^T z)^{1/2}$, and $\|x\| = (\|y\|^2 + \|z\|^2)^{1/2}$.

The vector functions $Y(t, y, z)$ and $G(t, y, z)$ in the domain

$$P = \{t \in R_+, \quad y \in D_y = \{y \colon \|y\| < H\}, \quad 0 \le \|z\| < \infty\} \tag{4.6.2}$$

satisfy the conditions for the existence and uniqueness of the solution of the Cauchy problem for the system (4.6.1). In addition, $Y(t, y, z)$ in the domain (4.6.2) satisfies the Lipschitz condition with respect to the variables y, z with a constant L and $Y(t, 0, z) = 0$, that is, the system (4.6.1) has the equilibrium $y = 0$ at $\mu = 0$.

For the system (4.6.1) assume that any solution $x(t, \mu)$ of this system is determined at all $t \ge 0$ for which $\|z(t, \mu)\| \le H$, $H = \text{const} > 0$, that is, the solution $x(t, \mu)$ is z-continuable.

Together with the generating system

$$\frac{dy}{dt} = Y(t, y, z), \quad y(t_0) = y_0,$$
$$\frac{dz}{dt} = Z(t, y, z), \quad z(t_0) = z_0, \tag{4.6.3}$$

we will consider the auxiliary function $v(t, x)$, $x = (y^T, z^T)^T$, defined in the domain (4.6.2). Recall the following definitions.

Definition 4.6.1 The function $v(t, y, z)$ is positive definite with respect to y in the domain (4.6.2) if and only if there exists a function a from the K-class, such that

$$a(\|y\|) \le v(t, y, z) \quad \text{at all} \quad (t, y, z) \in P.$$

Definition 4.6.2 The function $v(t, y, z)$ is decrescent with respect to y in the domain P, if and only if there exists a function b from the K-class such that

$$v(t, y, z) \le b(\|y\|) \quad \text{at all} \quad (t, y, z) \in P.$$

Definition 4.6.3 The state of equilibrium $x = 0$ of the system (4.6.1) is (y, μ)-stable, if for any $\varepsilon > 0$, $t_0 \in R_+$ there exists $\eta(\varepsilon) > 0$ and $\mu_0(\varepsilon) > 0$, not depending on t_0, such that for an arbitrary solution $x(t, t_0, x_0)$ of the system (4.6.1) the condition $\|y(t, t_0, x_0)\| < \varepsilon$ is satisfied at all $t \geq t_0$ as soon as $\|x_0\| < \eta$ and $\mu < \mu_0$.

The mean $\Theta(t_0, y_0, z_0)$ is less than zero in the set $E(v^* = 0)$ with respect to the variables y, if for any numbers η and ε, $0 < \eta < \varepsilon < H$, there exist positive numbers $r(\eta, \varepsilon)$ and $\delta(\eta, \varepsilon)$ such that $\Theta(t_0, y_0, z_0) < -\delta$ at $\eta \leq \|y_0\| \leq \varepsilon$, $\rho(y_0, E(v^* = 0)) < r(\eta, \varepsilon)$, $0 \leq \|z_0\| < \infty$. Here the continuous function $v^*(y)$ is determined at $y \in D_y$.

The following statement is correct.

Theorem 4.6.1 *Let the following conditions be satisfied in the domain (4.6.2):*

(1) *there exists a y-positive definite function $v(t, x)$, decrescent with respect to y and such that*

$$Dv(t, x)|_{(4.6.3)} \leq v^*(y) \leq 0;$$

(2) *there exist summable functions $M(t)$, $F(t)$, $N(t)$, constants M_0, F_0, and N_0, and a function $\chi(\beta) \in K$, such that the following inequalities hold:*

$$\varphi(t, x) \leq N(t), \quad \int_{t_1}^{t_2} N(t)\, dt \leq N_0(t_2 - t_1)$$

at $y \in D_y \setminus E(v^* = 0)$, $0 \leq \|z\| < \infty$, $t \in J$, and

$$\|G(t, x)\| \leq M(t), \quad \int_{t_1}^{t_2} M(t)\, dt \leq M_0(t_2 - t_1),$$

$$|\varphi(t, x') - \varphi(t, x'')| \leq \chi(\|x' - x''\|)F(t),$$

$$\int_{t_1}^{t_2} F(t)\, dt \leq F_0(t_2 - t_1)$$

in the domain (4.6.2) on any finite interval $[t_1, t_2]$;

(3) *uniformly with respect to t_0, x_0 there exists a mean $\Theta(t_0, y_0, z_0)$;*

(4) *the mean $\Theta(t_0, y_0, z_0)$ is less than zero in the set $E(v^* = 0)$ with respect to variables y.*

Then the solution $x = 0$ of the system (4.6.3) is (y, μ)-stable.

Proof Let $\varepsilon \in (0, H)$ and $t_0 \in R_+$ be specified. Assume that the conditions of Theorem 4.6.1 are satisfied. For the function $v(t, x)$ in view of condition (1) there exist functions $a \in K$ and $b \in K$ such that in the domain (4.6.2) the following inequality holds:

$$a(\|y\|) \leq v(t, x) \leq b(\|y\|). \qquad (4.6.4)$$

For all points of the moving surface $v(t, x) = a(\varepsilon/2)$ in view of the inequality (4.6.4) obtain

$$b^{-1}(a(\varepsilon/2)) \leq \|y\| \leq \varepsilon/2 \qquad (4.6.5)$$

for all $t \in R_+$. Let $\eta(\varepsilon) = b^{-1}(a(\varepsilon/2))$ and consider the solution $x(t, t_0, x_0)$ of the system (4.6.1) at $\|x_0\| < \eta(\varepsilon)$. Assume that it left the domain $\|y\| < \eta(\varepsilon)$ and at some point of time t_0' crossed the surface $v(t, x) = a(\varepsilon/2)$ in a point x_0'. For this point the inequality (4.6.5) is correct, and in view of condition (4) of the theorem there exists $r(\eta, \varepsilon/2)$ and $\delta(\eta, \varepsilon/2)$ such that one of the following conditions is satisfied:

$$\rho(y_0', E(v^* = 0)) > r(\eta, \varepsilon/2), \quad \Theta(t_0', y_0', z_0') < -\delta(\eta, \varepsilon/2).$$

Consider the following properties of the solution $x(t, t_0, x_0)$.

(a) Let the following conditions be satisfied at a point τ

$$v(\tau, x(\tau)) = a(\varepsilon/2), \quad \rho(y(\tau), E(v^* = 0)) \geq r(\eta, \varepsilon/2).$$

Consider the behavior of the function $v(t, x)$ along the solution $x(t, \tau, x(\tau)) = (y(t, \tau, x(\tau)), z(t, \tau, x(\tau)))$:

$$v(t, x(t)) \leq v(\tau, x(\tau)) + \int_\tau^t v^*(y(t)) \, dt + \mu \int_\tau^t \varphi(t, x(t)) \, dt. \qquad (4.6.6)$$

In this case, at $\mu < \mu_0' = \gamma/2N_0$ ($\gamma = \inf_{x \in Q} |v^*(y)|$, $Q = \{y: \rho(y, E(v^* = 0)) > r/2, \eta \leq \|y\| \leq \varepsilon/2\}$) for all $t > \tau$, for which the following conditions are satisfied:

$$\|y(t)\| \geq \eta, \quad \rho(y(t), E(v^* = 0)) > r/2,$$

obtain

$$v(t, x(t)) \leq a(\varepsilon/2) - \frac{\gamma}{2}(t - \tau). \qquad (4.6.7)$$

The function $v(t, x(t))$ is not increscent at $\mu < \mu_0'$, which means that for the solution $x(t, \tau, x(\tau))$ in view of the inequality (4.6.5) the condition $\|y(t)\| \leq \varepsilon/2$ will be satisfied, at least until the inequality

$$\rho(y(t), E(v^* = 0)) > r/2$$

is violated.

(b) Let the following conditions be satisfied at a point τ:

$$v(\tau, x(\tau)) = a(\varepsilon/2), \quad \rho(y(\tau), E(v^* = 0)) < r(\eta, \varepsilon/2).$$

In this case, under condition (4) of the theorem, $\Theta(\tau, y(\tau), z(\tau)) < -\delta$. Neglect the first integral in the inequality (4.6.6) and estimate the last one. For this purpose, choose a time interval l and the values μ_1 and μ_2 so that at $\mu < \mu_2$ on the interval $[\tau, \tau + 2l]$ for the solution $x(t, \tau, x(\tau))$ the condition $\|y(t, \tau, x(\tau))\| < \varepsilon$ will be satisfied. Choosing $\mu < \mu_0'' = \min(\mu_1, \mu_2)$, for the second integral from the inequality (4.6.6) at $t > \tau + l$ obtain the estimate

$$\int_\tau^t \varphi(t, x(t))\, dt \leq -\frac{\delta}{4}(t - \tau).$$

This integral, at least from the point of time $\tau + l$, becomes negative, which means that the solution $x(t, \tau, x(\tau))$ will return into the domain bounded by the surface $v(t, x) = a(\varepsilon/2)$.

As is clear from properties (a) and (b), for the solution $x(t, t_0, x_0)$ of the system (4.6.1) the condition $\|y(t, t_0, x_0)\| < \varepsilon$ will hold for all $t \geq t_0$, which proves the theorem.

Obviously, for the study of stability with respect to a part of variables it is also possible to use the perturbed Lyapunov function

$$v(t, x, \mu) = v_0(t, x) + u(t, x, \mu),$$

where the perturbation u at small values of μ may be sufficiently small.

4.7 Applications

4.7.1 Analysis of two weakly connected oscillators

Consider the system of two weakly connected oscillators

$$\begin{aligned}
\dot{x}_1 &= x_2, & \dot{x}_2 &= -\omega_1^2 x_1 + \mu\omega_1^2 x_1 x_3 x_4 \cos \nu_1 t, \\
\dot{x}_3 &= x_4, & \dot{x}_4 &= -\omega_2^2 x_3 + \mu\omega_2^2 x_1 x_2 x_3 \cos \nu_2 t.
\end{aligned} \tag{4.7.1}$$

The degenerate system corresponding to (4.7.1) is stable, which is determined by the existence of the Lyapunov function

$$v_0 = \omega_1^2 x_1^2 + x_2^2 + \omega_2^2 x_3^2 + x_4^2,$$

whose derivative in view of the degenerate system is identically zero. Hence it follows that the properties of stability or instability of the system (4.7.1) are determined by the sign of the mean $\Theta_0(t_0, x_0)$. The mean $\Theta_0(t_0, x_0)$ for the system (4.7.1) is determined by the expression

$$\Theta_0(t_0, x_0) = \Theta_0^{(1)}(t_0, x_0) + \Theta_0^{(2)}(t_0, x_0),$$

where

$$\Theta_0^i(t_0, x_0) = \lim_{T \to \infty} \frac{1}{T} \int_{t_0}^{t_0+T} \varphi^{(i)}(t, \overline{x}(t)) dt, \tag{4.7.2}$$

$$\varphi^{(i)}(t, \overline{x}(t)) = 2\omega_i^2 \overline{x}_1 \overline{x}_2 \overline{x}_3 \overline{x}_4 \cos \nu_i t$$

and $\overline{x}_i$, $i = 1, 2, 3, 4$, is a solution of the degenerate system corresponding to (4.7.1) under the initial condition $\overline{x}_i(0) = x_i^0$.

Denoting $\alpha = x_2^0(\omega_1 x_1^0)^{-1}$, $\beta = x_4^0(\omega_2 x_3^0)^{-1}$ and performing the necessary transformations, obtain

$$\Theta_0^i(t_0, x_0) = \frac{1}{8}\omega_1\omega_2\omega_i^2 x_1^{02} x_3^{02} \lim_{T \to \infty} \frac{1}{T} \int_{t_0}^{t_0+T} \{[4\alpha\beta + (\alpha^2 - 1)(\beta^2 - 1)]$$

$$\times [\cos(\gamma_1 - \nu_i t) + \cos(\gamma_1 + \nu_i t)] \tag{4.7.3}$$
$$+ [4\alpha\beta - (\alpha^2 - 1)(\beta^2 - 1)] [\cos(\gamma_2 - \nu_i t) + \cos(\gamma_2 + \nu_i t)]$$
$$+ 2[\beta(\alpha^2 - 1) - \alpha(\beta^2 - 1)] [\sin(\gamma_1 - \nu_i t) + \sin(\gamma_1 + \nu_i t)]$$
$$+ 2[\beta(\alpha^2 - 1) + \alpha(\beta^2 - 1)] [\sin(\gamma_2 - \nu_i t) + \sin(\gamma_2 + \nu_i t)]\}dt,$$

where $\gamma_1 = 2(\omega_1 - \omega_2)(t - t_0)$, $\gamma_2 = 2(\omega_1 + \omega_2)(t - t_0)$.

The mean $\Theta_0^{(i)}(t_0, x_0)$ is not identically zero in an arbitrarily small neighborhood of the point $x = 0$, except for the singular point, when $\nu_i = 2|\omega_1 - \omega_2|$ or $\nu_i = 2(\omega_1 + \omega_2)$.

Consider the first case in more detail. Let $\nu_i = 2|\omega_1 - \omega_2|$, assuming that $\omega_1 \neq \omega_2$, so that $\nu_i \neq 0$. Denoting $2(\omega_1 - \omega_2)t_0 = \gamma$, reduce the mean $\Theta_0^{(i)}(t_0, x_0)$ to the form

$$\Theta_0^i(t_0, x_0) = \frac{1}{8}\omega_1\omega_2\omega_i^2 x_1^{02} x_3^{02} \frac{1}{\cos\gamma}[\alpha(\beta\cos\gamma - \sin\gamma - 1) +$$

$$+ \beta(\sin\gamma + 1) + \cos\gamma] \times$$
$$\times [\alpha(\beta\cos\gamma - \sin\gamma + 1) + \beta(\sin\gamma - 1) + \cos\gamma].$$

The mean $\Theta_0(t_0, x_0)$ is nonzero when $\Theta_0^{(1)}(t_0, x_0) \neq 0$, or $\Theta_0^{(2)}(t_0, x_0) \neq 0$, or $\Theta_0^{(1)}(t_0, x_0) \neq 0$ and $\Theta_0^{(2)}(t_0, x_0) \neq 0$ simultaneously. It is easy to see that in all cases the mean $\Theta_0(t_0, x_0)$ is alternating in an arbitrary indefinitely small neighborhood of the stationary point.

Under the specific choice of ν_i, the system (4.7.1) satisfies the conditions of Theorem 4.2.2 on the instability of the equilibrium. To the system (4.7.1)

Theorem 4.3.1 is applicable. Determine the quantities included into the statement of the theorem. For simplicity of calculations we assume that the norm of vector $x \in R^n$ is specified by the expression $\|x\| = \max_i\{|x_i|\}$.

Then, since

$$\varphi(t, x) = 2x_1 x_2 x_3 x_4 (\omega_1^2 \cos \nu_1 t + \omega_2^2 \cos \nu_2 t), \qquad (4.7.4)$$

for the values of x, contained in the ring domain $\rho < \|x\| < \varepsilon$, where $\rho > 0$ and $\varepsilon > 0$ $(\rho < \varepsilon)$ are arbitrary constants, $t \in I$, one can assume that $\varphi_0(\varepsilon) = 2\varepsilon^4(\omega_1^2 + \omega_2^2)$, $\varphi_0(\bar{\varepsilon}) = 2\bar{\varepsilon}^4(\omega_1^2 + \omega_2^2)$.

Estimate the difference $\varphi(t, x') - \varphi(t, x'')$, $x', x'' \in B_\varepsilon$, $t \in I$:

$$
\begin{aligned}
|\varphi(t, x') - \varphi(t, x'')| &= |(x_1' x_2' x_3' x_4' - x_1'' x_2'' x_3'' x_4'')2(\omega_1^2 \cos \nu_1 t + \omega_2^2 \cos \nu_2 t) \\
&\leq (|(x_1' - x_1'')x_2' x_3' x_4'| + |x_1''(x_2' - x_2'')x_3' x_4'| \\
&\quad + |x_1'' x_2''(x_3' - x_3'')x_4'| + |x_1'' x_2'' x_3''(x_4' - x_4'')|) \cdot 2|\omega_1^2 \cos \nu_1 t + \omega_2^2 \cos \nu_2 t| \\
&< 4\varepsilon^3 \|x' - x''\| \cdot 2|\omega_1^2 \cos \nu_1 t + \omega_2^2 \cos \nu_2 t|.
\end{aligned}
$$

Taking into account condition (4) of Theorem 4.3.1, obtain

$$\chi(\alpha) = 4\varepsilon^3\alpha, \quad F_0 = 2(\omega_1^2 + \omega_2^2). \qquad (4.7.5)$$

For $(t, x) \in J \times B_\varepsilon$ $\|g(t,x)\| = \max\{\omega_1^2|x_1 x_3 x_4 \cos \nu_1 t|, \ \omega_2^2|x_1 x_2 x_3 \cos \nu_2 t|\}$ holds, hence it follows that $M_0 = \varepsilon^3 \max\{\omega_1^2, \omega_2^2\}$.

Specify a number $\varepsilon > 0$ and introduce a number $\bar{\varepsilon} > 0$ so that the condition $\bar{\varepsilon} < \varepsilon$ would be satisfied. Consider the surface of the level of the Lyapunov function $v_0(x) = \omega_1^2 x_1^2 + x_2^2 + \omega_2^2 x_3^2 + x_4^2 = w_{\bar{\varepsilon}}$.

For this surface to lie inside the ε-neighborhood of the origin of coordinates, it is obviously sufficient to choose

$$w_{\bar{\varepsilon}} = \bar{\varepsilon}^2 \min\{1, \omega_1^2, \omega_2^2\}. \qquad (4.7.6)$$

Similarly,

$$w_\varepsilon = \varepsilon^2 \min\{1, \omega_1^2, \omega_2^2\}. \qquad (4.7.7)$$

Determine η as the radius of the η-neighborhood of the stationary point, nested in the surface $v_0(x) = w_{\bar{\varepsilon}}$. For the point $x \in B_\eta$ obtain $v_0(x) \leq \eta^2(2 + \omega_1^2 + \omega_2^2) = \bar{\varepsilon}^2 \min\{1, \omega_1^2, \omega_2^2\}$, whence

$$\eta = \bar{\varepsilon} \frac{\min\{1, \omega_1^2, \omega_2^2\}}{\sqrt{2 + \omega_1^2 + \omega_2^2}}. \qquad (4.7.8)$$

Estimate the function $\varkappa(t)$ in the relation

$$\frac{1}{t - t_0} \int_{t_0}^{t} \varphi(t, \bar{x}(t))dt = \Theta_0(t_0, x_0) + \varkappa(t),$$

when ν_i do not satisfy the resonance conditions, and in the event when the resonance conditions are satisfied.

In the first case $\Theta_0^{(i)}(t_0, x_0) = 0$. Hence, taking into account (4.7.3), it follows that

$$|\varkappa_i(t)| = \frac{1}{t - t_0} \left| \int_{t_0}^{t} 2\omega_i^2 \overline{x_1 x_2 x_3 x_4} \cos \nu_i t \, dt \right|$$

$$< \frac{1}{t - t_0} \cdot \frac{\omega_1 \omega_2}{4} \omega_i^2 x_1^{02} x_3^{02} (\alpha^2 + 1)(\beta^2 + 1) d_i$$

$$< \frac{1}{t - t_0} \cdot \frac{\omega_i^2}{4\omega_1 \omega_2} \eta^4 (1 + \omega_2^2) d_i,$$

where

$$d_i = \frac{1}{|2(\omega_1 - \omega_2) - \nu_i|} + \frac{1}{|2(\omega_1 - \omega_2) + \nu_i|} +$$

$$+ \frac{1}{|2(\omega_1 + \omega_2) - \nu_i|} + \frac{1}{2(\omega_1 + \omega_2) + \nu_i}.$$

If the value ν_i is resonance, then $\Theta_0^{(i)}(t_0, x_0) \neq 0$ and obtain a similar estimate for $\varkappa_i(t)$, but the expression for d_i does not contain a summand whose denominator is zero at the specified value of ν_i.

Thus, at $t \in [t_0 + l, t_0 + 2l]$ the condition $|\varkappa(t)| < \delta/4$ will be satisfied, when l is determined by the inequality

$$l \geq \frac{\eta^4}{\delta} \frac{(1 + \omega_1^2)(1 + \omega_2^2)}{\omega_1 \omega_2} (\omega_1^2 d_1 + \omega_2^2 d_2). \tag{4.7.9}$$

Find μ_1 by the formula

$$\mu_1 = (\varepsilon - \bar{\varepsilon}) \left[M_0 \cdot 2l e^{2lN} \right]^{-1}, \tag{4.7.10}$$

where $N = \max\{\omega_1^2, \omega_2^2\}$. Choose μ_2 so small that the inequality $\chi(\mu_2 M_0 \cdot 2l e^{2lN}) F_0 \leq \delta/4$ would hold. From the last expression, taking into account (4.7.5), obtain

$$\mu_2 = \frac{\delta}{16\varepsilon^3 F_0} [M_0 \cdot 2l e^{2lN}]^{-1}. \tag{4.7.11}$$

Now determine $\mu_0 = \min\{\mu_1, \mu_2\}$.

Applying Theorem 4.3.1 to the system (4.7.1), one can formulate the following statement.

Corollary 4.7.1 Let at a specified number $\varepsilon > 0$ and some $\bar{\varepsilon}$, $0 < \bar{\varepsilon} < \varepsilon$, the quantities $\varphi_0(\varepsilon)$, $\varphi_0(\bar{\varepsilon})$, w_ε, $w_{\bar{\varepsilon}}$, η, l, μ_0 be determined by the expressions (4.7.4)–(4.7.11). Then at $\mu < \mu_0$ the solution $x(t; t_0, x_0)$ of the system (4.7.1), which began at a point $x_0 \in E_\Theta^-(t_0) \cap B_\eta$, will not leave the domain $\|x\| < \varepsilon$ on the time interval $t \in [t_0, t_0 + T]$, where $T = 2l + l\delta[\varphi_0(\bar{\varepsilon})]^{-1} + (w_\varepsilon -$

$w_{\bar{\varepsilon}})[\mu\varphi_0(\varepsilon)]^{-1}$ and the quantity δ at the specified t_0, x_0 is determined by the inequality $0 < \delta < -\Theta_0(t_0, x_0)$.

Let in the system (4.7.1) $\omega_1 = 1$, $\omega_2 = 2$, $\nu_1 = \nu_2 = 2|\omega_1 - \omega_2| = 2$. Having determined $\varphi_0(\varepsilon)$, $\varphi_0(\bar{\varepsilon})$, w_ε, $w_{\bar{\varepsilon}}$, η, l, μ_0 from the expressions (4.7.4)–(4.7.11), we arrive at the following statement.

Corollary 4.7.2 The solution of the system (4.7.1), satisfying the condition $x_0 \in E_\Theta^-(t_0)$ at the initial point, with $\|x_0\| < \dfrac{\varepsilon}{\sqrt{14}}$, at $\mu < \mu_0$ will not leave the ε-neighborhood of the point $x = 0$ on an interval $t \in [t_0, t_0 + T]$, where

$$T = 2l + \frac{2l\delta}{5\varepsilon^4} + \frac{1}{20\mu\varepsilon^2}, \quad \mu_0 = \min\left\{(2 - \sqrt{2})\varepsilon, \frac{\delta}{80\varepsilon^3}\right\}\frac{e^{-8l}}{16l\varepsilon^3},$$

$$l \geq \frac{125}{784}\frac{\varepsilon^4}{\delta},$$

and the quantity δ is determined from the above relations.

In the event when neither of the resonance conditions is satisfied in the system (4.7.1), the mean $\Theta_0(t_0, x_0)$ is identically zero, and it is necessary to use the perturbed Lyapunov function $v(t, x, \mu)$ to solve the question of stability of the system (4.7.1). Let

$$v(t, x, \mu) = v_0(t, x) + \mu v_1(t, x).$$

Differentiate the perturbed Lyapunov function in view of the equations of the system (4.7.1) (here it is taken into account that a nonperturbed system is neutrally stable)

$$\frac{dv}{dt} = \mu\left(\frac{\partial v_1}{\partial t} + \frac{\partial v_1}{\partial x}f(t, x) + \varphi(t, x)\right) + \mu^2\frac{\partial v_1}{\partial x}g(t, x).$$

Determine $v_1(t, x)$ as a solution of the linear equation in first-order partial derivatives

$$\frac{\partial v_1}{\partial t} + \frac{\partial v_1}{\partial x}f(t, x) = -\varphi(t, x). \tag{4.7.12}$$

Since the characteristics of the equation (4.7.12) are the integral curves of the nonperturbed system (4.2.3), by integrating the equation (4.7.12) with the initial conditions $v_1(0, x_0) = 0$ obtain the function $v_1(t, x)$ which is equal to zero on the initial set $t = 0$, x_0:

$$v_1(t, x) = -\int_0^t \varphi(\tau, \bar{x}(\tau))\, d\tau.$$

For the system (4.7.1) the equation (4.7.12) takes the form

$$\frac{\partial v_1}{\partial t} + x_2\frac{\partial v_1}{\partial x_1} - \omega_1^2 x_1\frac{\partial v_1}{\partial x_2} + x_4\frac{\partial v_1}{\partial x_3} - \omega_2^2 x_3\frac{\partial v_1}{\partial x_4} = -\varphi(t, x),$$

whence

$$v_1(t, x) = -2 \int\limits_0^t \overline{x}_1 \overline{x}_2 \overline{x}_3 \overline{x}_4 (\omega_1^2 \cos \nu_1 \tau + \omega_2^2 \cos \nu_2 \tau) \, d\tau.$$

The system (4.7.1) with the constructed perturbed Lyapunov function $v(t, x, \mu) = v_0(x) + \mu v_1(t, x)$ satisfies the conditions of theorems from Section 4.2 on the stability and instability of a stationary point. Therefore, the stability of the system is determined by the sign of the mean $\Theta_1(t_0, x_0) = \Theta_1^{(1)}(t_0, x_0) + \Theta_1^{(2)}(t_0, x_0)$, where

$$\Theta_1^{(1)}(t_0, x_0) = \lim_{T \to \infty} \frac{1}{T} \int\limits_{t_0}^{t_0+T} \omega_1^2 \frac{\partial v_1}{\partial x_2} \overline{x}_1 \overline{x}_3 \overline{x}_4 \cos \nu_1 t \, dt,$$

$$\Theta_1^{(2)}(t_0, x_0) = \lim_{T \to \infty} \frac{1}{T} \int\limits_{t_0}^{t_0+T} \omega_2^2 \frac{\partial v_1}{\partial x_4} \overline{x}_1 \overline{x}_2 \overline{x}_3 \cos \nu_2 t \, dt.$$

The final form of the function and the means is not given here due to their awkwardness.

Like in the treatment of the mean $\Theta_0(t_0, x_0)$, in the case under consideration it turns out that in an arbitrary indefinitely small neighborhood of the point $x = 0$ the mean $\Theta_1(t_0, x_0)$ is alternating. In the same manner as for the mean $\Theta_0(t_0, x_0)$, obtain that the solution of the system (4.7.1) is unstable at those values of ν_1 and ν_2, at which $\Theta_1(t_0, x_0) \neq 0$, that is, the resonance occurs. Here the resonance values will be those of ω_1, ω_2, ν_1, and ν_2, for which the following relations hold:

$$\omega_1 = 2\omega_2, \quad 2\omega_1 = \omega_2, \quad \nu_1 = \omega_1, \quad \nu_2 = \omega_2, \quad \nu_1 = 2\omega_2,$$
$$\nu_2 = 2\omega_1, \quad \nu_1 = \omega_1 + 2\omega_2, \quad \nu_1 = |\omega_1 - 2\omega_2|, \quad \nu_2 = 2\omega_1 + \omega_2,$$
$$\nu_2 = |2\omega_1 - \omega_2|, \quad \nu_1 + \nu_2 = 2\omega_1, \quad \nu_1 + \nu_2 = 2\omega_2, \quad |\nu_1 - \nu_2| = 2\omega_1,$$
$$|\nu_1 - \nu_2| = 2\omega_2, \quad \nu_1 + \nu_2 = 4\omega_1, \quad \nu_1 + \nu_2 = 4\omega_2, \quad |\nu_1 - \nu_2| = 4\omega_1,$$
$$|\nu_1 - \nu_2| = 4\omega_2, \quad \nu_1 - \nu_2 = \pm 2\omega_1 \pm 4\omega_2, \quad \nu_1 - \nu_2 = \pm 4\omega_1 \pm 2\omega_2,$$
$$\nu_1 + \nu_2 = |2\omega_1 \pm 4\omega_2|, \quad \nu_1 + \nu_2 = |4\omega_1 \pm 2\omega_2|.$$

4.7.2 System of n oscillators

In the space R^{2n} consider a system n of weakly connected oscillators

$$\dot{x}_{2i-1} = x_{2i},$$
$$\dot{x}_{2i} = -\omega_i^2 x_1 + \mu \omega_i^2 \cos \nu_i t \prod_{j=1, j \neq 2i}^{2n} x_j, \quad i = 1, 2, \ldots, n. \tag{4.7.13}$$

In a similar manner as was done for the system (4.7.1), determine that the degenerate system corresponding to (4.7.13) is stable, since the derivative, in view of the degenerate system of the Lyapunov function

$$v_0 = \sum_{i=1}^{n} (\omega_i^2 x_{2i-1}^2 + x_{2i}^2),$$

is identically zero. Therefore, the properties of stability or instability of the system (4.7.13) are determined by the sign of the mean $\Theta_0(t_0, x_0)$, which is specified by the expression

$$\Theta_0(t_0, x_0) = \sum_{i=1}^{n} \Theta_0^{(i)}(t_0, x_0),$$

where

$$\Theta_0^i(t_0, x_0) = \lim_{T \to \infty} \frac{2\omega_i^2}{T} \int_{t_0}^{t_0+T} \prod_{k=1}^{2n} \overline{x}_k \cos \nu_i t \, dt$$

and $\overline{x}_i$, $i = 1, 2, \ldots, 2n$, is a solution of the degenerate system corresponding to (4.7.13) at the initial condition $\overline{x}_i(0) = x_i^0$.

After necessary transformations, transform the last expression as follows:

$$\Theta_0^i(t_0, x_0) = \frac{\omega_i^2}{2^{n-1}} \prod_{k=1}^{n} \frac{\omega_k^2 x_{2k-1}^{02} + x_{2k}^{02}}{\omega_k}$$

$$\times \lim_{T \to \infty} \frac{1}{T} \int_{t_0}^{t_0+T} \prod_{k=1}^{h} \sin[2\omega_k(t - t_0) + \psi_k] \cos \nu_i t \, dt$$

where $\mathrm{tg}\, \psi_k = 2 \dfrac{\omega_k x_{2k-1}^0 x_{2k}^0}{-\omega_k^2 x_{2k-1}^{02} + x_{2k}^{02}}$.

Let P denote an integer n-vector whose components p_i take on the values ± 1, $\overline{\Omega} = (\omega_1, \omega_2, \ldots, \omega_n) \in R^n$, $\Psi = (\psi_1, \ldots, \psi_n) \in R^n$, $\langle P, \overline{\Omega} \rangle = p_1 \omega_1 + \ldots + p_n \omega_n$. Upon necessary transformations, for $\Theta_0^{(i)}(t_0, x_0)$ obtain the expression

$$\Theta_0^{(i)}(t_0, x_0) = \frac{\omega_i^2}{2^{2n-1}} \prod_{k=1}^{n} \left(\omega_k x_{2k-1}^{02} + \frac{x_{2k}^{02}}{\omega_k} \right) \times$$

$$\times \lim_{T \to \infty} \frac{1}{T} \int_{t_0}^{t_0+T} \sum_k \sin[2\langle P_k, \overline{\Omega} \rangle (t - t_0) + \langle P_k, \Psi \pm \nu_i t \rangle] \, dt.$$

The limit in the right-hand part is not zero if and only if at least for one value of k

$$2|\langle P_k, \overline{\Omega} \rangle| = \nu_i, \quad i = 1, 2, \ldots, n. \tag{4.7.14}$$

Here $\Theta_0^{(i)}(t_0, x_0)$ takes the form

$$\Theta_0^{(i)}(t_0, x_0) = \frac{\omega_i^2}{2^{2n-1}} \prod_{k=1}^{n} \left(\omega_k x_{2k-1}^{02} + \frac{x_{2k}^{02}}{\omega_k} \right) \sum_{k'} \sin(\langle P_{k'}, \Psi - 2\overline{\Omega} t_0 \rangle), \tag{4.7.15}$$

where the summation is applied to those indices k' for which (4.7.14) holds. It is obvious that $\Theta_0^{(i)}(t_0, x_0)$ determined in such manner is alternating in an arbitrary indefinitely small neighborhood of the point $x = 0$.

It is easy to show that in view of the degenerate system, the following relation holds:

$$\frac{d}{dt}\Theta_0^{(i)}(t, \bar{x}) = 0.$$

Therefore, on the solutions of the degenerate system corresponding to (4.7.13), the mean remains constant

$$\Theta_0^{(i)}(t, \bar{x}(t; t_0, x_0)) = \Theta_0^{(i)}(t_0, x_0).$$

Thus, the system (4.7.13) satisfies the conditions of Theorem 4.2.2, and at resonance values ν_i determined by the relation (4.7.14) the origin of coordinates is an unstable equilibrium. In particular, at $n = 1$ the resonance value will be $\nu = 2\omega$; at $n = 2 - \nu_i = 2(\omega_1 + \omega_2)$, $\nu_i = 2|\omega_1 - \omega_2|$; at $n = 3$ we obtain four resonance values ν_i: $2(\omega_1 + \omega_2 + \omega_3)$, $2|\omega_1 + \omega_2 - \omega_3|$, $2|\omega_1 - \omega_2 + \omega_3|$, $2|-\omega_1 + \omega_2 + \omega_3|$.

Let us use Theorem 4.3.1 to estimate the time interval on which the solution $x(t; t_0, x_0)$ will not leave the ε-neighborhood of the point $x = 0$. For this purpose it is necessary to estimate $\varphi(t, x)$ and determine the values φ_0, σ, and η contained in the statement of the theorem. As it was done before, determine the norm of vector $x \in R^n$ by the expression $\|x\| = \max_i\{|x_i|\}$. Then for the values of x contained in the ring area $\rho < \|x\| < \varepsilon$, where $\rho, \varepsilon = \text{const} > 0$, obtain

$$|\varphi(t, x)| = 2\left|\left(\prod_{i=1}^{2n} x_i\right)\sum_{i=1}^{n}\omega_i^2 \cos\nu_i t\right| < 2\varepsilon^{2n}\sum_{i=1}^{n}\omega_i^2.$$

Thus, one can assume that $\varphi_0 = 2\varepsilon^{2n}\sum_{i=1}^{n}\omega_i^2$.

To determine $\sigma(\varepsilon)$ and $\eta(\varepsilon)$, consider the surface of the level of the Lyapunov function

$$v_0(x) = \sum_{i=1}^{n}(\omega_i^2 x_{2i-1}^2 + x_{2i}^2) = w_0. \tag{4.7.16}$$

Obviously, the surface (4.7.16) will lie inside the ε-neighborhood of the point $x = 0$ if we assume that

$$w_0 = \varepsilon^2 \min_i\{1, \omega_i^2\}. \tag{4.7.17}$$

Determine $\sigma(\varepsilon)$ by the inequality $\sigma(\varepsilon) < w_0$, assuming that

$$\sigma(\varepsilon) = (1 - \lambda^2)w_0 = (1 - \lambda^2)\varepsilon^2 \min_i\{1, \omega_i^2\}, \quad 0 < \lambda < 1.$$

Determine the constant $\eta(\varepsilon)$ as the radius of the η-neighborhood of the point $x = 0$ lying inside the surface $v_0(x) = w_0 - \sigma$. For the point $x \in B_\eta$ obtain

$$v_0(x) = \sum_{i=1}^{n}(\omega_i^2 x_{2i-1}^2 + x_{2i}^2) \leq \eta^2 \sum_{i=1}^{n}(\omega_i^2 + 1) = \lambda^2 \varepsilon^2 \min_i\{1, \omega_i^2\},$$

whence

$$\eta(\varepsilon) = \varepsilon \, \frac{\lambda \min_i \{1, \omega_i^2\}}{\sqrt{n + \sum\limits_{i=1}^{n} \omega_i^2}}. \tag{4.7.18}$$

Thus, we arrive at the following statement.

Corollary 4.7.3 Solutions of the system (4.7.13) which at the initial point t_0 satisfy the inequality $\|x_0\| < \eta$, where η is determined by the expression (4.7.18), for all $t \in [t_0, t_0 + T]$ will not leave the domain $\|x\| < \varepsilon$, if T is determined by the expression

$$T = (1 - \lambda^2) \min_i \{1, \omega_i^2\} \left(2\mu\varepsilon^{2n-2} \sum_{i=1}^{n} \omega_i^2 \right)^{-1}. \tag{4.7.19}$$

Since in the system (4.7.13) the mean is alternating in sign in an arbitrary indefinitely small neighborhood of the origin of coordinates, we can apply to it Theorem 4.3.2 on the stability on a finite interval for solutions beginning in the domain $E_{\Theta}^-(t_0) = \{x_0 \in B_\eta : \Theta_0(t_0, x_0) < 0\}$. In the same manner as was done for the system of two oscillators, obtain

$$M_0 = \varepsilon^{2n-1} \max_i \{\omega_i^2\}, \quad \chi(\alpha) = 2n\varepsilon^{2n-1}\alpha, \quad F_0 = 2\sum_{i=1}^{n} \omega_i^2,$$

$$a = \min_i \{1, \omega_i\}, \quad \omega_\varepsilon = \varepsilon^2 a^2, \quad \omega_{\bar\varepsilon} = \bar\varepsilon^2 a^2, \quad \eta = \bar\varepsilon a \left(n + \sum_{i=1}^{n} \omega_i^2 \right)^{-\frac{1}{2}},$$

$$N = \max_i \{1, \omega_i^2\}, \quad \varphi_0(\varepsilon) = 2\varepsilon^{2n} \sum_{i=1}^{n} \omega_i^2, \quad \varphi_0(\bar\varepsilon) = 2\bar\varepsilon^{2n} \sum_{i=1}^{n} \omega_i^2,$$

$$l \geq \frac{\eta^{2n}}{2^{2n-3}\delta} \prod_{i=1}^{n} \frac{1 + \omega_i^2}{\omega_i} \cdot \sum_{i=1}^{n} \omega_i^2 d_i,$$

$$\mu_0 = \min \left\{ \varepsilon - \bar\varepsilon, \; \frac{\delta}{8n\varepsilon^{2n-1}F_0} \right\} \cdot (M_0 \cdot 2le^{2lN})^{-1} \tag{4.7.20}$$

where

$$d_i = \sum_k \left(\frac{1}{|2\langle P_k, \overline{\Omega}\rangle - \nu_i|} + \frac{1}{|2\langle P_k, \overline{\Omega}\rangle + \nu_i|} \right).$$

If the value of ν_i satisfies the resonance conditions (4.7.14), then the expression for d_i does not contain a summand whose denominator is equal to zero at the specified value of ν_i.

Applying Theorem 4.3.2 to the system (4.7.13), one can formulate the following statement.

Corollary 4.7.4 Let at the specified number $\varepsilon > 0$ and some $\bar{\varepsilon}$, $0 < \bar{\varepsilon} < \varepsilon$, the values of ω_ε, $\omega_{\bar{\varepsilon}}$, l, μ_0, φ_0, and η be determined by the expressions (4.7.20). Then at $\mu < \mu_0$ the solution $x = x(t; t_0, x_0)$ of the system (4.7.13), emanating from the point $x_0 \in E_{\bar{\Theta}}^-(t)$ such that $\|x_0\| < \eta$, will not leave the domain $\|x\| < \varepsilon$ on the time interval $t \in [t_0, t_0 + T]$, where

$$T = 2l + l\delta \left(2\bar{\varepsilon}^{2n} \sum_{i=1}^{n} \omega_i^2 \right)^{-1} + a(\varepsilon^2 - \bar{\varepsilon}^2) \left(2\mu\varepsilon^{2n} \sum_{i=1}^{n} \omega_i^2 \right)^{-1},$$

and the value δ at the specified t_0 and x_0 is determined from the inequality $0 < \delta < -\Theta_0(t_0, x_0)$.

Now revert to Corollary 4.7.3. Let T^* denote the point of time at which the solution reaches the surface $S(\varepsilon) = \{x : \|x\| = \varepsilon\}$. Excluding λ from (4.7.18) and (4.7.19), for T^* obtain the estimate

$$T^* \geq T = \frac{\varepsilon^2 \min_i \{1, \omega_i^2\} - \eta^2 \left(n + \sum_{i=1}^{n} \omega_i^2 \right)}{2\mu\varepsilon^{2n} \sum_{i=1}^{n} \omega_i^2}, \tag{4.7.21}$$

whence it follows that at a constant ε with a decrease of the η-neighborhood of the origin of coordinates the value of T approaches the limit

$$T' = \frac{\min_i \{1, \omega_i^2\}}{2\mu\varepsilon^{2n-2} \sum_{i=1}^{n} \omega_i^2}.$$

If at a constant η ε is unlimitedly increased, then at $n \neq 1$ T vanishes.

Thus, the expression (4.7.21) gives a substantially low value of T, since it would be natural to expect that in the first case T will unboundedly increase, and in the second case it would at least not decrease. This shortcoming is eliminated by the application of Theorem 4.3.3, which more completely takes into account the information on the change of perturbations $g(t, x)$ in the ε-neighborhood of the origin of coordinates.

Since for the system (4.7.13)

$$\varphi(t, x) = 2 \prod_{i=1}^{2n} x_i \cdot \sum_{i=1}^{n} \omega_i^2 \cos \nu_i t,$$

then

$$\psi(\zeta) = 2\zeta^{2n} \sum_{i=1}^{n} \omega_i^2. \tag{4.7.22}$$

For the function $\varkappa(\zeta)$, in view of (4.7.17), obtain the expression

$$\varkappa(\zeta) = \zeta^2 \min_i \{1, \omega_i^2\}. \tag{4.7.23}$$

Determine the constant $\eta(\varepsilon)$ as the radius of the η-neighborhood of the point $x = 0$ lying inside the surface S_0. Assuming that $\varepsilon_0 = \lambda\varepsilon$, $0 < \lambda < 1$, for $\eta(\varepsilon)$ obtain the expression

$$\eta(\varepsilon) = \lambda\varepsilon \min_i\{1, \omega_i\}\left(n + \sum_{i=1}^n \omega_i^2\right)^{-\frac{1}{2}}. \tag{4.7.24}$$

Substituting (4.7.22) and (4.7.23) into (4.3.3) and integrating, we arrive at the following statement.

Corollary 4.7.5 Solutions of the system (4.7.13), which at the initial point of time t_0 satisfy the inequality $\|x_0\| < \eta$, where the value is determined by the expression (4.7.24), for all $t \in [t_0, t_0 + T]$ will not leave the domain $\|x\| < \varepsilon$ if at $n \neq 1$ T is determined by the expression

$$T = \frac{\min_i\{1, \omega_i^2\}}{2(n-1)\mu\varepsilon^{2n-2}\sum_{i=1}^n \omega_i^2}\left(\frac{1}{\lambda^{2n-2}} - 1\right) \tag{4.7.25}$$

and at $n = 1$

$$T = \frac{\min\{1, \omega^2\}}{\mu\omega^2}\ln\frac{1}{\lambda},$$

where $0 < \lambda < 1$.

Excluding λ from (4.7.25), using (4.7.24), at $n \neq 1$ obtain

$$T = \frac{\min_i\{1, \omega_i^2\}}{2(n-1)\mu\sum_{i=1}^n \omega_i^2}\left(\frac{\min_i\{1, \omega_i^{2n-2}\}}{\eta^{2n-2}\left(n + \sum_{i=1}^n \omega_i^2\right)^{n-1}} - \frac{1}{\varepsilon^{2n-2}}\right),$$

whence it follows that at a constant ε at a decrease in the η-neighborhood of the origin of coordinates the value $T \to \infty$, and at an unlimited increase of ε and a constant η the value of T approaches the limit

$$T'' = \frac{\min_i\{1, \omega_i^2\}}{2(n-1)\mu\eta^{2n-2}\sum_{i=1}^n \omega_i^2\left(n + \sum_{i=1}^n \omega_i^2\right)^{n-1}}.$$

Thus, Corollary 4.7.5 gives a much better estimate of T than Corollary 4.7.3.

Since in the system (4.7.13) the mean $\Theta_0(t_0, x_0)$ is alternating in an arbitrarily small neighborhood of the origin of coordinates, we can apply to it Theorem 4.3.4 on the stability on a finite interval for solutions beginning in the domain $E_\Theta^-(t_0 \cap B_\eta)$. Here the values $\eta(\bar{\varepsilon})$, μ_0, l, and the functions $\psi(\zeta)$ and $\varkappa(\zeta)$ contained in the statement of the theorem are determined by the expressions (4.7.20), (4.7.22), and (4.7.23), respectively. Find ε_0 from the relation

$$\varepsilon_0 = \left(\bar{\varepsilon}^2 - \frac{l\delta\mu}{\min_i\{1, \omega_i^2\}}\right)^{\frac{1}{2}}. \tag{4.7.26}$$

Applying Theorem 4.3.4, obtain the following statement.

Corollary 4.7.6 Let at the specified $\varepsilon > 0$ and some $\bar{\varepsilon}$, $0 < \bar{\varepsilon} < \varepsilon$, the values $\eta(\bar{\varepsilon})$, μ_0, l be determined by the expressions (4.7.20), and the functions $\psi(\zeta)$, $\varkappa(\zeta)$, and ε_0 by the expressions (4.7.22), (4.7.23), and (4.7.26), respectively. Then at $\mu < \mu_0$ the solution $x = x(t; t_0, x_0)$ of the system (4.7.13), emanating from the point $x_0 \in E_{\Theta}^{-}(t_0)$ such that $\|x_0\| < \eta$, will not leave the domain $\|x\| < \varepsilon$ on the time interval $t \in [t_0, t_0 + T]$, where

$$T = 2l + \frac{\min\limits_{i}\{1, \omega_i^2\}}{2(n-1)\mu \sum_{i=1}^{n} \omega_i^2} \left(\frac{1}{\varepsilon_0^{2n-2}} - \frac{1}{\varepsilon^{2n-2}} \right)$$

and the value of δ at specified t_0, x_0 is determined from the inequality $0 < \delta < -\Theta_0(t_0, x_0)$.

In conclusion we note that Theorem 4.3.3, like the theorem of the averaging method, gives the estimate of closeness of solutions on a time interval of the order $1/\mu$. On the other hand, this theorem is close to theorems on stability on a finite interval, which may be obtained on the basis of the comparison principle. An advantage of Theorem 4.3.3 is the simplicity of definition of the functions $\psi(\zeta)$ and $\varkappa$.

4.8 Comments and References

The application of the averaging technique in the investigation of real-world processes dates back to the works of Euler [1], Lagrange [1], Poincaré [2], and other founders of the mathematical science. In the study of stability of solutions of nonautonomous systems, the averaging technique is applied in many works (see, e.g., Bogolyubov and Mitropolsky [1], Starzhinsky [1], Roso [1], Martynyuk [4], Sanders and Verhulst [1], and others).

This chapter contains some results of the analysis of stability of nonlinear systems with a small parameter on the basis of the combination of the ideas of the method of Lyapunov functions and the averaging principle of nonlinear mechanics).

4.2. This section is based on the results of the articles of Martynyuk and Kosolapov [1] and Kosolapov [3].

4.3. Here the results of the article of Chernetskaya [2] were used. The cited results adjoin the investigation of Khapaev [1] (Theorem 4.3.1) and the results obtained by other authors using a similar technique of analysis (see Anashkin [1, 2]).

4.4. Here some results obtained by Karimzhanov [1], Karimzhanov and Kosolapov [1], Martynyuk and Karimzhanov [1, 2], and Martynyuk and Chernetskaya [1] are used.

4.5. In the works of Martynyuk [1–3], it was for the first time shown that a method of estimating the stability of motion of large-scale systems could be applied, which used special means of the derivative of an auxiliary function along solutions of independent subsystems. This approach was developed in the works of Kosolapov [1, 2, 4] and others (see, e.g., Karimzhanov and Kosolapov [1] and others). In this section, the articles of Martynyuk and Kosolapov [1, 2] were used.

4.6. This section is based on the results obtained by Kosolapov (see [1, 3]).

4.7. This section is based on the results obtained by Chernetskaya (see [1, 2]).

Chapter 5

Stability of Systems in Banach Spaces

5.1 Introductory Remarks

In this chapter the results of the analysis of μ-stability and boundedness of solutions of equations in Banach spaces are given. Those equations describe the class of hybrid systems with weakly interacting subsystems.

Section 5.2 contains some results from the theory of semigroups, which are required for further treatment.

In Section 5.3, the problem of stability of systems in Banach spaces with weakly interacting subsystems is formulated.

Section 5.4 contains the description of the general method of solution of the posed problem. The application of the matrix-valued Lyapunov function is discussed and the main theorems of that method are formulated for equations in a Banach space.

In Section 5.5, the vector Lyapunov function is applied and the results of the analysis of stability of a system in Banach spaces are given.

In Section 5.6, a matrix-valued function is applied to the analysis of μ-stability of a two-component hybrid system. The case of nonasymptotic stability of isolated subsystems is considered.

The concluding section contains bibliographic data and some remarks on further investigation in this line.

5.2 Preliminary Results

Let $\mathbb{X}$ or $\mathbb{Z}$ denote a Banach space, and let a linear operator A be defined in the domain $\mathcal{D}(A) \subset \mathbb{X}$ with its rank in $\mathbb{Z}$, that is, $A\colon \mathcal{D}(A) \to \mathbb{Z}$. Assume that $\mathcal{D}(A)$ is a dense linear subspace $\mathbb{X}$. The operator A is closed if its graph $Gr(A) = \{(x, Ax) \in \mathbb{X} \times \mathbb{Z}\colon x \in \mathcal{D}(A)\}$ is a closed subset in the product $\mathbb{X} \times \mathbb{Z}$. For the specified linear mapping $A\colon \mathcal{D}(A) \to \mathbb{Z}$, $\mathcal{D}(A) \subset \mathbb{X}$, its norm is determined by the expression

$$\|A\| = \sup\{\|Ax\|\colon \|x\| = 1\},$$

and $\rho(A)$ is a resolvent set of the operator A.

Assume that some physical process is described by the linear differential equation

$$\frac{dx}{dt} = Ax, \tag{5.2.1}$$

$$x(0) = x_0 \in \mathcal{D}(A) \tag{5.2.2}$$

at all $t \in \mathbb{R}_+$. The abstract Cauchy problem (5.2.1) and (5.2.2) is defined correctly if $\rho(A) \neq \varnothing$ and for any $x_0 \in \mathcal{D}(A)$ there exists a unique solution $x \colon [0, \infty) \to \mathcal{D}(A)$ in the space $C^1([0, \infty), \mathbb{X})$.

The family $(Q(t))_{t \geq 0}$ of bounded linear operators acting in a Banach space $\mathbb{X}$ is a strictly continuous semigroup of bounded linear operators (C_0-semigroup) if the following conditions are satisfied:

(a) $Q(0) = I$, I is an identical operator on $\mathbb{X}$;

(b) $Q(t)Q(s) = Q(t + s)$ at all $t, s \geq 0$;

(c) $\lim\limits_{t \downarrow 0} \|Q(t)x - x\| = 0$ at all $x \in \mathbb{X}$.

The infinitisemal generator of the semigroup $(Q(t))_{t \geq 0}$ is a linear operator A with the domain of definition

$$\mathcal{D}(A) = \left\{ x \in \mathbb{X} \colon \lim_{t \downarrow 0} \frac{1}{t}(Q(t)x - x) \quad \text{exists} \right\}$$

in the form

$$Ax = \lim_{t \downarrow 0} \frac{1}{t}(Q(t)x - x), \quad x \in \mathcal{D}(A).$$

Along with the problem (5.2.1) and (5.2.2) consider the nonlinear abstract Cauchy problem

$$\frac{dx}{dt} = A(x(t)), \tag{5.2.3}$$

$$x(0) = x_0 \in \mathcal{D}(A), \tag{5.2.4}$$

where $A \colon \mathcal{D}(A) \to \mathbb{X}$ is a nonlinear mapping. Assume that the solution $x(t)$ of this problem is determined correctly and exists on $\mathbb{R}_+ = [0, \infty)$.

Let $\mathcal{C}$ be a subset of the Banach space $\mathbb{X}$. The family $(Q(t))_{t \geq 0}$ of operators mapping $\mathcal{C}$ into $\mathcal{C}$ is a nonlinear subgroup on $\mathcal{C}$ if the mapping $Q(t)x$ is continuous with respect to (t, x) on the product $\mathbb{R}_+ \times \mathcal{C}$, $Q(0)x = x$ and $Q(t + s)x = Q(t) \times Q(s)x$ for any fixed $x \in \mathcal{C}$ at $(t, s) \in \mathbb{R}_+$.

A nonlinear semigroup $Q(t)$ is quasicontracting if there exists a number $w \in \mathbb{R}$ such that $\|Q(t)x - Q(t)y\| \leq e^{wt}\|x - y\|$ at all $t \in \mathbb{R}_+$ and all $x, y \in \mathcal{C}$.

5.3 Statement of the Problem

Assume that for the equation (5.2.3) a linear (or nonlinear) semigroup $Q(t)$ is defined on a subspace $\mathcal{C} \in \mathbb{X}$. Let the point $0 \in \operatorname{int} \mathcal{C}$ and $Q(t)$ permit the trivial solution $Q(t)x = 0$ at all $t \in \mathbb{R}_+$ and $x = 0$.

Definition 5.3.1 The trivial solution $Q(t)x = 0$ of the equation (5.2.3) is stable if for any $\varepsilon > 0$ there exists $\delta = \delta(\varepsilon) > 0$ such that $\|Q(t)x\| < \varepsilon$ at all $t \in \mathbb{R}_+$, as soon as $\|x\| < \delta$ at $x \in \mathcal{C}$.

Definitions of other types of stability of the trivial solution $Q(t)x = 0$ of the equation (5.2.3) are introduced in the same way as it was done for the finite-dimensional case in view of Definition 5.3.1.

Now consider the nonlinear equations

$$\frac{dx_i}{dt} = f_i(x_i), \quad i = 1, 2, \ldots, m, \tag{5.3.1}$$

and assume that the corresponding abstract Cauchy problem is correctly defined. Let the semigroup $Q_i(t)$ be defined on $\mathcal{C}_i \subset \mathbb{Z}_i$ and the point $0 \in \operatorname{int} \mathcal{C}_i$ at any $i = 1, 2, \ldots, m$. The domain $\mathcal{D}(f_i)$ is assumed to be dense in $\mathcal{C}_i$ and the functions f_i are generators of the semigroups $Q_i(t)$.

Using the operators $g_i(x, \mu)$, $i = 1, 2, \ldots, m$ ($\mu \in M = (0, 1]$ is a small positive parameter) defined on $\mathcal{D}(g_i) \times M \subset \mathbb{X}$ and having the rank in $\mathbb{Z}_i$, combine the equations (5.3.1) into the system

$$\frac{dx_i}{dt} = f_i(x_i) + g_i(x, \mu), \quad i = 1, 2, \ldots, m. \tag{5.3.2}$$

In particular, the operators $g_i(x, \mu)$ may have the form

(A) $g_i(x, \mu) = \sum\limits_{s=1}^{\infty} \mu^s G_{is}(x_1, \ldots, x_m)$, $\quad i = 1, 2, \ldots, m$,

(B) $g_i(x, \mu) = \sum\limits_{s=1}^{N-1} \mu^s G_{is}(x_1, \ldots, x_m)$, $\quad i = 1, 2, \ldots, m$,

(C) $g_i(x, \mu) = \mu G_i(x_1, \ldots, x_m)$, $\quad i = 1, 2, \ldots, m$.

The operators G_{is} are assumed to be defined on $\mathcal{D}(G_{is}) \subset \mathbb{X}$ (on $\mathcal{D}(G_i) \subset \mathbb{X}$) and having the rank in $\mathbb{Z}_i$. Here $x_i \in \mathbb{Z}_i$ and the hypervector $x^{\mathrm{T}} = (x_1, \ldots, x_m)$ is a point in the product of spaces

$$\mathbb{X} = \prod_{i=1}^{m} \mathbb{Z}_i$$

with the norm $\|x\| = \sum_{i=1}^{m} \|x_i\|_i$. The system of equations (5.3.2) is equivalent
to the equation

$$\frac{dx}{dt} = f(x) + g(x,\mu) \overset{\Delta}{=} A(x,\mu),$$

$$x(0) = x_0 \in \mathcal{D}(f + g(x,\mu)),$$

$$(5.3.3)$$

where $f^{\mathrm{T}}(x) = (f_1(x_1),\ldots,f_m(x_m))$, $g^{\mathrm{T}}(x,\mu) = (g_1(x,\mu),\ldots,g_m(x,\mu))$.

The system (5.3.2) is a hybrid system with weakly interacting subsystems
(5.3.1). Note that

$$\mathcal{D}(f + g(x,\mu)) = \mathcal{D}(f) \cap \mathcal{D}(g(x,\mu)) =$$
$$= \mathcal{D}(f) \cap \mathcal{D}(g_1(\mu)) \cap \mathcal{D}(g_2(\mu)) \cap \ldots \cap \mathcal{D}(g_m(\mu)).$$

In addition, it is assumed that the equation (5.3.3) is correctly defined, the
vector function $f(x) + g(x,\mu)$ generates the semigroup $Q(t)$, and the domain
$\mathcal{D}_0 = \mathcal{D}(f(x) + g(x,\mu)) \cap \mathcal{D}(f_s) \cap \mathcal{D}(f(x) + g(x,\mu))_s$ is dense in X.

Our objective is to find the method for the analysis of μ-stability of the zero
solution of the system (5.3.2) on the basis of the generalized direct Lyapunov
method.

5.4 Generalized Direct Lyapunov Method

Along with the system (5.3.2) consider the two-index system of functions

$$U(x) = [u_{ij}(x)], \quad i,j = 1,2,\ldots,s, \quad s \le m, \quad (5.4.1)$$

with the elements $u_{ii}: Z_i \to \mathbb{R}_+$ and $u_{ij}: Z_i \times Z_j \to \mathbb{R}$ at all $i \neq j$. Let
$\theta \in \mathbb{R}_+^s$, $\theta_i > 0$, and the function

$$v(x,\theta) = \theta^{\mathrm{T}} U(x)\theta \quad (5.4.2)$$

satisfy the conditions:

(1) there exists a neighborhood $W \in \mathbb{X}$ of the point $0 \in \operatorname{int} \mathcal{C}$, such that
$v\colon W \to \mathbb{R}_+$;

(2) the function $v(x,\theta)$ is continuous with respect to $x \in W$ and $v(x,\theta) = 0$,
if and only if $x = 0$;

(3) there exists a limit

$$\lim_{t \to 0^+} \sup \frac{v(Q(t)x,\theta) - v(x,\theta)}{t} = Dv(x(t),\theta)$$

along the path $x(t) = Q(t)x_0$ of the system (5.3.3).

The function (5.4.2) will be called the Lyapunov function for the system (5.3.2) in Banach space, if it satisfies conditions $(1)-(3)$ and solves the problem of stability (instability) of the zero solution $Q(t)x = 0$ of the system (5.3.2).

Note that the elements $u_{ii}(x)$, $i = 1, 2, \ldots, s$, of the matrix function (5.4.1) are constructed on the basis of the equations (5.3.1) or their linear approximation, and the elements $u_{ij}(x_i, x_j)$ at $(i \neq j) \in [1, s]$ are constructed in view of the connection operators $g_i(x, \mu)$ or on the basis of consideration of the pairs of subsystems

$$\frac{dx_i}{dt} = f_i(x_i),$$

$$\frac{dx_j}{dt} = f_j(x_j)$$

at $(i \neq j) \in [1, s]$. Generally, this approach simplifies the problem of construction of an appropriate Lyapunov function (functional) for the system (5.3.2) in Banach space.

Let us cite the main theorems of the generalized direct Lyapunov method for the system (5.3.2).

Theorem 5.4.1 *If at some natural $s \leq m$ the function $v(x, \theta)$, $\theta \in \mathbb{R}^s_+$, is a Lyapunov function and there exists a comparison function φ_1 belonging to K-class, such that $v(x, \theta) \geq \varphi_1(\|x\|)$ in the neighborhood W of the point $0 \in \operatorname{int} C$, and if $Dv(x, \theta)|_{(5.3.3)} \leq 0$ at all $x \in W$ and $\mu < \mu^* \in M$, then the trivial solution $Q(t)x = 0$ of the system (5.3.2) is μ-stable.*

Theorem 5.4.2 *If at some natural $s \leq m$ for the function $v(x, \theta)$, $\theta \in \mathbb{R}^s_+$, there exist three comparison functions $\varphi_1, \varphi_2, \varphi_3$ of class K, such that $\varphi_1(\|x\|) \leq v(x, \theta) \leq \varphi_2(\|x\|)$ in the neighborhood W of the point $0 \in \operatorname{int} C$, and $Dv(x, \theta)|_{(5.3.3)} \leq -\varphi_3(\|x\|)$ at all $x \in W$ and $\mu < \mu^* \in M$, then the trivial solution $Q(t)x = 0$ of the system (5.3.2) is uniformly asymptotically μ-stable.*

Theorem 5.4.3 *If in the conditions of Theorem 5.4.2 $W = C = \mathbb{X}$ and the comparison function φ_1 belongs to KR-class, then the trivial solution $Q(t)x = 0$ of the system (5.3.2) is globally uniformly asymptotically μ-stable.*

Theorem 5.4.4 *If in the conditions of Theorem 5.4.2 the comparison functions φ_2, φ_3 belong to K-class and have the same order of growth, there exists a positive constant Δ_1 and an integer p such that*

$$\Delta_1 \|x\|^p \leq v(x, \theta) \leq \varphi_2(\|x\|),$$

then the trivial solution $Q(t)x = 0$ of the system (5.3.2) is exponentially μ-stable.

Theorem 5.4.5 *If in the conditions of Theorem 5.4.2 $W = C = \mathbb{X}$ and in the conditions of Theorem 5.4.4 the comparison functions φ_2, φ_3 belong to*

KR-class and have the same order of growth, then the trivial solution $Q(t)x = 0$ of the system (5.3.2) is globally exponentially μ-stable.

Theorem 5.4.6 *Let at some natural $s \leq m$ for the function $v(x,\theta)$, $\theta \in \mathbb{R}^s_+$, there exist a comparison function φ from K-class, such that $-Dv(x,\theta)|_{(5.3.3)} \geq \varphi(\|x\|)$ in the neighborhood $W \subset C$ of the point $0 \in \text{int}\, C$ at any $\mu \in M$. If in any neighborhood $N \subset C$ of the point $0 \in \text{int}\, C$ there exists at least a single point $x_0 \in N$ at which $v(x_0,\theta) < 0$, then the trivial solution $Q(t)x = 0$ of the system (5.3.2) is μ-unstable.*

Theorem 5.4.7 *Let $C = \mathbb{X}$ and $S = \{x \in \mathbb{X}\colon \|x\| \geq r\}$, where r may be sufficiently large. If at some natural $s \leq m$ for the function $v(x,\theta)\colon S \to \mathbb{R}_+$, $\theta \in \mathbb{R}^s_+$, there exist two comparison functions φ_1, φ_2 from KR-class, such that*

$$\varphi_1(\|x\|) \leq v(x,\theta) \leq \varphi_2(\|x\|)$$

at all $x \in S$, and if $Dv(x,\theta)|_{(5.3.3)} \leq 0$ at all $x \in S$ and $\mu < \mu^* \in M$, then the path $Q(t)x_0$ of the system (5.3.2) is uniformly μ-bounded.

Theorem 5.4.8 *Let the conditions of Theorem 5.4.7 be satisfied and let there exist a comparison function φ_3 from K-class, such that $Dv(x,\theta)|_{(5.3.3)} \leq -\varphi_3(\|x\|)$ at all $x \in S$ and $\mu < \mu^* \in M$. Then the path $Q(t)x_0$ of the system (5.3.2) is uniformly ultimately μ-bounded.*

The constructive application of Theorems 5.4.1–5.4.8 is connected with the solution of the problem of construction of the function (5.4.2) with properties (1) and (2) and the calculation of its full derivative $Dv(x,\theta)$ along the path $x(t) = Q(t)x_0$ of the system (5.3.3). In a general case, the second problem is quite intricate. In some cases its solution may be simplified. Precisely, if the semigroup $Q(t)$ is a C_0-semigroup or a quasicontracting semigroup on a Hilbert space or a uniformly convex Banach space, then the infinitesimal generator A_s of the semigroup $Q(t)$ exists on a set $\mathcal{D}(A_s)$ which is dense in C. In such a case, the calculation of $Dv(x,\theta)|_{(5.3.3)}$ is simplified.

The pair $(Q(t), v)$ is permissible for the problem (5.2.2), if v is a Lyapunov function, the infinitesimal generator A_s of the semigroup $Q(t)$ is defined on the set $\mathcal{D}_0 \subset \mathcal{D}(A_s)$ dense in C, and, in addition, there exists a function ∇v defined on $(W \cap \mathcal{D}_0) \times X$, with its values in $\mathbb{R}$, such that

(a) $v(y) - v(x) \leq \nabla v(x, y - x) + o(\|y - x\|)$ at all $x, y \in \mathcal{D}_0$ and

(b) at each fixed x the operator $\nabla v(x, \theta, h)$ is bounded and linear with respect to $h \in \mathbb{X}$.

Theorem 5.4.9 (see Michel and Miller [1, pp. 143–144]) *Let for the system (5.3.3) there exist a permissible pair $(Q(t), v)$ and a comparison function φ belonging to K-class, such that $\nabla v(x, \theta, A_s x) \leq -\varphi(\|x\|)$ at all $x \in \mathcal{D}_0 \cap W$. Then $Dv(x,\theta)|_{(5.3.3)} \leq -\varphi(\|x\|)$ at all $x \in W$.*

Proof Let $x \in \mathcal{D}_0 \cap W$. Then, according to the definition of the function $Dv(x, \theta)$, obtain

$$Dv(x, \theta) = \limsup_{t \to 0^+} \frac{v(\varphi(t)x, \theta) - v(x, \theta)}{t}$$

$$\leq \limsup_{t \to 0^+} \frac{\nabla v(x, Q(t)x - x, \theta) + o(\|Q(t)x - x\|)}{t}$$

$$= \limsup_{t \to 0^+} \nabla v(x, (Q(t)x - x)/t, \theta)$$

$$= \nabla v(x, A_s x, \theta) \leq -\varphi(\|x\|).$$

Now assume that $x \notin \mathcal{D}_0 \cap W$. Choose a sequence $\{x_n\}$ in $\mathcal{D}_0$ so that $x_n \to x$ at $n \to +\infty$. Since any element x_n belongs to $\mathcal{D}_0$, at all $t \in \mathbb{R}_+$

$$v(Q(t)x_n, \theta) - x(x_n, \theta) \leq -\int_0^t \varphi(\|Q(s)x_n\|) \, ds.$$

The continuity of all functions contained in the above inequality implies that

$$v(Q(t)x, \theta) - x(x, \theta) \leq -\int_0^t \varphi(\|Q(s)x\|) \, ds.$$

Hence, obtain

$$\limsup_{t \to 0^+} \frac{v(Q(t)x, \theta) - v(x, \theta)}{t} \leq \limsup_{t \to 0^+} \left(-\frac{1}{t}\right) \int_0^t \varphi(\|Q(s)x\|) \, ds = -\varphi(\|x\|).$$

Thus, at all $x \in W$ obtain the estimate $Dv(x, \theta) \leq \leq -\varphi(\|x\|)$. Theorem 5.4.9 is proved.

Note that along with the function (5.4.2) in some cases it makes sense to apply the vector function

$$V(x, B, \theta) = BU(x)\theta, \quad \theta \in \mathbb{R}_+^s, \tag{5.4.3}$$

where B is an $(s \times s)$-constant matrix. The vector function $V(x, B, \theta)$ has the scalar functions $v_i(x, B, \theta)$, $i = 1, 2, \ldots, s$, as its components. If in the expression (5.4.1) $u_{ij}(\cdot) = 0$ at all $(i \neq j) \in [1, s]$, then $U(x)$ is a vector function, that is, $U(x) = \text{diag}\,[u_{11}(x), \ldots, u_{ss}(x)]$.

5.5 μ-Stability of Motion of Weakly Connected Systems

In this section we will consider the system (5.3.2) with the subsystems (5.3.1). The dynamic properties of the zero solution $Q_i(t)x_i = 0$ of the subsystem (5.3.1) will be characterized as follows.

Let for each subsystem from the collection (5.3.1) there exist a semigroup $Q_i(t)$ and a scalar function $v_i(x_i)$ such that the pair $(Q_i(t), v_i)$ is permissible.

Assumption 5.5.1 An isolated subsystem from the collection (5.3.1) permits the property A, if for the pair $(Q_i(t), v_i)$ there exist functions $\psi_{i1}, \psi_{i2}, \psi_{i3}$ from K-class and constants $\Delta_i > 0$ and β_i such that:

(1) $\psi_{i1}(\|x_i\|) \leq v_i(x_i) \leq \psi_{i2}(\|x_i\|)$ at all $x_i \in Z_i$, such that $\|x_i\| < \Delta_i$, and

(2) $\nabla v_i(x_i, {}^s f_i(x_i)) \leq \beta_i \psi_{i3}(\|x_i\|)$ at all $x_i \in \mathcal{D}({}^s f_i)$ such that $\|x_i\| < \Delta_i$.

Here ${}^s f_i$ is an infinitesimal generator of the semigroup $Q_i(t)$.

Assumption 5.5.2 An isolated subsystem from the collection (5.3.1) permits the property B, if it has the property A at $\Delta_i = +\infty$ and comparison functions ψ_{i1}, ψ_{i2} from KR-class.

Assumption 5.5.3 The operator of connection $g_i(x, \mu)$ between the subsystems (5.3.1) satisfies the property C, if at the specified permissible pair $(Q_i(t), v_i)$ there exist comparison functions ψ_{i3} from K-class and constants $b_{ij}(\mu)$, $i, j = 1, 2, \ldots, m$, such that

$$\nabla v_i(x_i, g_i(x, \mu)) \leq \psi_{i3}^{1/2}(\|x_i\|) \sum_{j=1}^{m} b_{ij}(\mu) \psi_{j3}^{1/2}(\|x_j\|) \qquad (5.5.1)$$

at all $x^{\mathrm{T}} = (x_1, \ldots, x_m) \in \mathcal{D}(f + g(x, \mu))$ and $\|x_i\| < \Delta_i$, $i = 1, 2, \ldots, m$.

For the class of systems in Banach space with the subsystems (5.3.1) and operators of connection between subsystems $g_i(x, \mu)$, satisfying the properties A and C, respectively, the following statement is correct.

Theorem 5.5.1 *Assume that for each subsystem of the system (5.3.2) in Banach space there exists a semigroup $Q_i(t)$ and a function $v_i(x_i)$, composing the permissible pair $(Q_i(t), v_i)$, and*

(1) *the isolated subsystems from the collection (5.3.1) permit the property A;*

(2) *the operators of connection $g_i(x, \mu)$ between the subsystems (5.3.1) permit the property C;*

(3) *there exist constants $\theta_i > 0$, $i = 1, 2, \ldots, m$, and a value of the parameter $\mu^* \in M$ such that the matrix $A(\mu) = [a_{ij}(\mu)]$ with the elements*

$$a_{ij}(\mu) = \begin{cases} \theta_i(\beta_i + b_{ii}(\mu)) & \text{at} \quad i = j, \\ \frac{1}{2}(\theta_i b_{ij}(\mu) + \theta_j b_{ji}(\mu)) & \text{at} \quad i \neq j \end{cases}$$

is negative definite at $\mu < \mu^*$.

Then the trivial solution of the system (5.3.2) is uniformly asymptotically μ-stable.

Proof Over the set $\Pi = \{x^{\mathrm{T}} = (x_1, \ldots, x_m) \colon \|x_i\| < \Delta_i$ at $i = 1, 2, \ldots, m\}$ consider the function

$$v(x, \theta) = U^*(x)\theta, \quad \theta \in \mathbb{R}_+^m, \tag{5.5.2}$$

where $U^*(x) = \mathrm{diag}[u_{11}(x_1), \ldots, u_{mm}(x_m)]$. According to the conditions of the theorem, the functions $u_{ii}(x_i) = v_i(x_i)$ together with the semigroup $Q_i(t)$ form a permissible pair for the i-th subsystem from the collection (5.3.1). It is obvious that $v(x, \theta)$ is a continuous function and $v(0, \theta) = 0$. Since $v_i(x_i, \theta)$ satisfies condition (1), then

$$\sum_{i=1}^{m} \theta_i \psi_{i1}(\|x_i\|) \le v(x, \theta) \le \sum_{i=1}^{m} \theta_i \psi_{i2}(\|x_i\|)$$

at all $x \in \Pi$. For the functions ψ_{i1}, ψ_{i2} from K-class one can find comparison functions ψ_1, ψ_2 belonging to the K-class, such that

$$\psi_1(\|x\|) \le v(x, \theta) \le \psi_2(\|x\|) \tag{5.5.3}$$

at all $x \in \Pi$, where

$$\psi_1(\|x\|) \le \sum_{i=1}^{m} \theta_i \psi_{i1}(\|x_i\|)$$

and

$$\psi_2(\|x\|) \ge \sum_{i=1}^{m} \theta_i \psi_{i2}(\|x_i\|).$$

For the values of $x \in W_0 \subset \mathbf{X}$ calculate the difference

$$v(x + h, \theta) - v(x, \theta) = \sum_{i=1}^{m} \theta_i \{v_i(x_i + h_i) - v_i(x_i)\}$$

$$\le \sum_{i=1}^{m} \theta_i \{\nabla v_i(x_i, h_i) + o(\|h_i\|)\} = \sum_{i=1}^{m} \theta_i \nabla v_i(x_i, h_i) + o(\|h\|).$$

Hence it follows that $\nabla v(x, \theta, h) = \sum_{i=1}^{m} \theta_i \nabla v_i(x_i, h_i)$. From the fact that $\nabla v_i(x_i, h_i)$ are continuous and linear with respect to h_i, it follows that $\nabla v(x, \theta, h)$ at each fixed $x \in \Pi$ is continuous and linear with respect to h.

In view of the above and Assumption 5.5.3, obtain

$$\nabla v(x, \theta, f(x) + g(x, \mu)) = \sum_{i=1}^{m} \theta_i \nabla v_i(x_i, f(x) + g(x, \mu))$$

$$= \sum_{i=1}^{m} \beta_i \nabla v_i(x_i, f(x)) + \sum_{i=1}^{m} \beta_i \nabla v_i(x_i, g(x, \mu))$$

$$\leq \sum_{i=1}^{m} \theta_i \left[\beta_i \psi_{i3}(\|x_i\|) + \psi_{i3}^{1/2}(\|x_i\|) \sum_{j=1}^{m} b_{ij}(\mu) \psi_{j3}^{1/2}(\|x_j\|) \right]$$

$$= u^{\mathrm{T}} A(\mu) u,$$

where the elements $a_{ij}(\mu)$ of the matrix $A(\mu) = [a_{ij}(\mu)]$ are the same as in condition (3) of Theorem 5.5.1, and the vector u is determined as follows:

$$u^{\mathrm{T}} = \left[\psi_{13}^{1/2}(\|x_1\|), \ldots, \psi_{m3}^{1/2}(\|x_m\|) \right].$$

Since at $\mu < \mu^* \in M$ the matrix $A(\mu)$ is negative definite, then

$$\nabla v(x, \theta, f(x) + g(x, \mu)) \leq u^{\mathrm{T}} A(\mu) u \leq \lambda_M(A) \|u\|^2,$$

where $\lambda_M(A) < 0$ at $\mu < \mu^*$. The fact that

$$\|u\|^2 = \sum_{i=1}^{m} \psi_{i3}(\|x_i\|) \geq \psi_3(\|x\|)$$

for some function ψ_3 from K-class implies the estimate

$$\nabla v(x, \theta, f(x) + g(x, \mu)) \leq \lambda_M(A) \psi_3(\|x\|)$$

at all $x \in \Pi \cap W_0$. Hence, according to Theorem 5.4.9, obtain the estimate

$$Dv(x, \theta)|_{(5.3.3)} \leq \lambda_M(A) \psi_3(\|x\|), \qquad (5.5.4)$$

which in view of Theorem 5.4.2 secures the uniform asymptotic μ-stability of the zero solution $Q(t)x = 0$ of the system (5.3.2).

Theorem 5.5.2 *Assume that for each subsystem of the system (5.3.2) in Banach space there exists a semigroup $Q_i(t)$ and a function $v_i(x_i)$, which form a permissible pair $(Q_i(t), v_i)$, and*

(1) *the isolated subsystems from the collection (5.3.1) permit the property B;*

(2) *under the specified functions $v_i(x_i)$ and the comparison functions ψ_{i3} from K-class there exist constants $b_{ij}^*(\mu)$, $i, j = 1, 2, \ldots, m$, such that the estimates*

$$\nabla v_i(x_i, g_i(x, \mu)) \leq \psi_{i3}^{1/2}(\|x_i\|) \sum_{j=1}^{m} b_{ij}^*(\mu) \psi_{j3}^{1/2}(\|x_j\|)$$

hold at all $x \in \mathcal{D}(f + g(x, \mu))$, where $x^{\mathrm{T}} = (x_1, \ldots, x_m) \in \mathbb{X}$;

(3) *there exist constants $\theta_i > 0$, $i = 1, 2, \ldots, m$, and a value of the parameter $\mu^* \in M$ such that the matrix $A(\mu) = [a_{ij}(\mu)]$ with the elements*

$$a_{ij}(\mu) = \begin{cases} \theta_i(\beta_i + b_{ii}^*(\mu)) & at \quad i = j, \\ \frac{1}{2}\left(\theta_i b_{ij}^*(\mu) + \theta_j b_{ji}^*(\mu)\right) & at \quad i \neq j \end{cases}$$

is negative definite at $\mu < \mu^*$.

Then the trivial solution of the system (5.3.2) is globally uniformly asymptotically μ-stable.

Proof Under condition (1) of Theorem 5.5.2 the function (5.5.2) is estimated by the comparison functions $\psi_1(\|x\|)$ and $\psi_2(\|x\|)$ belonging to KR-class, and the estimate (5.5.3) holds at all $x \in \mathbb{X} = \prod_{i=1}^{m} \mathbb{X}_i$. Under condition (2) of Theorem 5.5.2, the estimate (5.5.4) takes the form

$$Dv(x, \theta)|_{(5.3.3)} \leq \lambda_M(A)\psi_3(\|x\|),$$

where $\psi_3(\|x\|) \leq \sum_{j=1}^{m} \psi_{j3}(\|x\|)$ at all $x \in \mathbb{X}$.

According to Theorem 5.4.3, the zero solution $x_i = Q_i(t)x_{i0} = 0$ of the system (5.3.2) is globally uniformly asymptotically μ-stable.

For the analysis of the exponential μ-stability of the system (5.3.2) we will need some assumptions on the functions $v_i(x_i)$ for the subsystems (5.3.1).

Assumption 5.5.4 An isolated subsystem from the collection (5.3.1) permits the property A*, if for the pair $(Q_i(t), v_i)$ there exist comparison functions ψ_{i2}, ψ_{i3} from K-class of the same order of growth, constants a_i, r_i, Δ_i and arbitrary constants β_i such that:

(1) $a_i\|x_i\|^{r_i} \leq v_i(x_i) \leq \psi_{i2}(\|x\|)$ at all $x_i \in \mathbb{Z}_i$ such that $\|x_i\| < \Delta_i$, and

(2) $\nabla v_i(x_i, {}^s f_i) \leq \beta_i \psi_{i3}(\|x_i\|)$ at all $x_i \in \mathcal{D}({}^s f_i)$ such that $\|x_i\| < \Delta_i$.

Assumption 5.5.5 An isolated subsystem from the collection (5.3.1) permits the property B* if it has the property A* at $\Delta_i = \infty$ and at comparison functions of the same order of growth ψ_{i2}, ψ_{i3}, belonging to KR-class.

Now prove the following statement.

Theorem 5.5.3 *Assume that for each subsystem of the system (5.3.2) a semigroup $Q_i(t)$ and a function $v_i(x_i)$ are constructed which form a permissible pair, and*

(1) *isolated subsystems from the collection (5.3.1) permit property A*;*

(2) *the operators of connection $g_i(x, \mu)$ between the subsystems (5.3.1) satisfy property C;*

(3) *there exist constants $\theta_i > 0$, $i = 1, 2, \ldots, m$, and a value of the parameter $\mu^* \in M$ such that the matrix $A(\mu) = [a_{ij}(\mu)]$, $i, j = 1, 2, \ldots, m$, from condition (3) of Theorem 5.5.1 is negative definite at $\mu < \mu^*$.*

Then the trivial solution of the system (5.3.2) is uniformly exponentially μ-stable.

Proof Like in the proof of Theorem 5.5.1, apply the function (5.5.2). Under condition (1) of Assumption 5.5.4 obtain the estimate for the function $v(x, \theta)$:

$$\min_i(\theta_i a_i) \sum_{i=1}^m \|x_i\|^{r_i} \leq v(x, \theta) \leq \psi_2(\|x\|), \tag{5.5.5}$$

where $\psi_2(\|x\|) \geq \sum_{j=1}^m \theta_j \psi_{j2}(\|x_j\|)$, ψ_2 belongs to K-class and has an inverse function $\psi_2^{-1}(\|x\|)$.

Under conditions (2) and (3) of Theorem 5.5.3 obtain

$$Dv(x, \theta)|_{(5.3.2)} \leq \lambda_M(A)\psi_3(\|x\|), \tag{5.5.6}$$

where $\psi_3(\|x\|) \leq \sum_{i=1}^m \psi_{i3}(\|x_i\|)$, $\lambda_M(A) < 0$ at $\mu < \mu^*$. Taking into account that the comparison functions $\psi_2(\|x\|)$ and $\psi_3(\|x\|)$ have the same order of growth at all $x \in \Pi = \{x^T = (x_1, \ldots, x_m): \|x_i\| < \Delta_i$ at all $i = 1, 2, \ldots, m\}$, transform the estimates (5.5.5) and (5.5.6). There exist constants k_1 and $k_2 > 0$ such that

$$k_1\psi_2(\|x\|) \leq \psi_3(\|x\|) \leq k_2\psi_2(\|x\|) \tag{5.5.7}$$

at all $x \in \Pi$.

Denote $a = \min_i(\theta_i a_i)$, $\|x\|^r = \sum_{i=1}^m \|x_i\|^{r_i}$, and let $k_1 = -\lambda_M(A)$. Then the estimates (5.5.5) and (5.5.6) take the form

$$a\|x\|^r \leq v(x, \theta) \leq \psi_2(\|x\|),$$
$$Dv(x, \theta)|_{(\cdot)} \leq -k_1 v(x, \theta)$$

at all $x \in \Pi$. Hence obtain

$$v(x(t), \theta) \leq v(x_0, \theta) \exp[-k_1(t - t_0)], \quad t \geq t_0.$$

Taking into account the inequality in the left-hand part of (5.5.7), obtain

$$\|x(t)\| \leq a^{-1/r}\psi_2^{1/r}(\|x_0\|) \exp\left[-\frac{k_1}{r}(t - t_0)\right] \tag{5.5.8}$$

at all $t \geq t_0$.

Denote $\lambda = \dfrac{k_1}{r}$ and at any $0 < \varepsilon < H$ choose $\delta(\varepsilon) = \psi^{-1}(a\varepsilon^r)$. Then at $\|x_0\| < \delta(\varepsilon)$ the estimate (5.5.8) implies that

$$\|x(t)\| \leq \varepsilon \exp[-\lambda(t - t_0)], \quad t \geq t_0.$$

Theorem 5.5.3 is proved.

Theorem 5.5.4 *Assume that for each subsystem of the system (5.3.2) there exists a semigroup $Q_i(t)$ and a function $v_i(x_i)$, which form a permissible pair, and*

(1) *isolated subsystems from the collection (5.3.1) permit the property B^*;*

(2) *conditions (2) and (3) of Theorem 5.5.2 are satisfied.*

Then the trivial solution of the system (5.3.2) is globally exponentially μ-stable.

Proof Under the conditions of Assumption 5.5.5 for the function $v(x, \theta)$ obtain the estimate

$$\min_i (\theta_i b_i) \sum_{i=1}^{m} \|x_i\|^{r_i} \le v(x, \theta) \le \psi_2(\|x\|), \tag{5.5.9}$$

where $b_i > 0$, $r_i > 0$ and $\psi_2(\|x\|)$ is a function from KR-class, which has an inverse $\psi_2(\|x\|) \ge \sum_{j=1}^{m} \theta_j \psi_{i2}(\|x_i\|)$. For the function $Dv(x, \theta)$ obtain

$$Dv(x, \theta)|_{(5.3.2)} \le \lambda_M(\Lambda)\psi_3(\|x\|),$$

where $\psi_3(\|x\|) \le \sum_{j=1}^{m} \psi_{i3}(\|x_i\|)$, $\lambda_M(A) < 0$ t $\mu < \mu^*$.

Similarly to the proof of Theorem 5.5.3 it is easy to obtain the estimate

$$\|x(t)\| \le b^{-1/r}\psi_2^{1/r}(\|x_0\|)\exp\left[-\frac{k_1}{r}(t - t_0)\right], \quad t \ge t_0.$$

For any $\alpha > 0$ calculate $K(\alpha) = b^{-1/r}\psi_2^{1/r}(\alpha)$. Here as soon as $\|x_0\| \le \alpha$, then $\|x(t)\| \le K(\alpha)\exp[-\lambda(t - t_0)]$, $t \ge t_0$, at any $x^T = (x_1, \ldots, x_m) \in \mathbb{X}$.
Theorem 5.5.4 is proved.

Now we will give the conditions for the μ-stability of the system (5.3.2) on the basis of the function (5.5.2) in which $\theta = (1, 1, \ldots, 1) \in \mathbb{R}_+^m$, and the constants $b_{ij} \ge 0$ at all $i \ne j$ and $\mu \le \mu^* \in M$. For this purpose, instead of the matrix $A(\mu)$ with the elements $a_{ij}(\mu)$, $i, j = 1, 2, \ldots, m$, consider a matrix $S(\mu)$ with the elements

$$s_{ij}(\mu) = \begin{cases} -(\beta_i + b_{ii}(\mu)) & \text{at} \quad i = j, \\ -b_{ij}(\mu) & \text{at} \quad i \ne j, \end{cases} \tag{5.5.10}$$

where β_i are constants from condition (2) of Assumption 5.5.1 and $b_{ij}(\mu)$ are constants from the estimate (5.5.1). Note that in further consideration of the properties of global stability in the expressions s_{ij} of the matrix $S^*(\mu)$ we will use the constants $b_{ij}^*(\mu) \ge 0$ at all $i \ne j$ and $\mu < \mu^* \in M$.

Consider the following statement.

Theorem 5.5.5 *Assume that for each subsystem of the system (5.3.2) there exist a semigroup $Q_i(t)$ and a function $v_i(x_i)$, which form a permissible pair $(Q_i(t), v_i)$, and, in addition:*

(1) *conditions (1) and (2) of Theorem 5.5.1 with the constants $b_{ij}(\mu) \geq 0$ at $i \neq j$ and all $\mu < \mu^* \in M$ are satisfied. If the main diagonal minors of the matrix $S(\mu)$ are positive at all $\mu < \mu^*$, then the trivial solution of the system (5.3.2) is uniformly asymptotically μ-stable;*

(2) *conditions (1) and (2) of Theorem 5.5.2 with the constants $b_{ij}^*(\mu) \geq 0$ at $i \neq j$ and all $\mu < \mu^* \in M$ are satisfied. If the main diagonal minors of the matrix $S^*(\mu)$ are positive at all $\mu < \mu^*$, then the trivial solution of the system (5.3.2) is globally uniformly asymptotically μ-stable;*

(3) *conditions (1) and (2) of Theorem 5.5.3 with the constants $b_{ij}(\mu) \geq 0$ at $i \neq j$ and all $\mu < \mu^* \in M$ are satisfied. If the main diagonal minors of the matrix $S(\mu)$ are positive at all $\mu < \mu^*$, then the trivial solution of the system (5.3.2) is exponentially μ-stable;*

(4) *conditions (1) and (2) of Theorem 5.5.4 with the constants $b_{ij}^*(\mu) \geq 0$ at $i \neq j$ and all $\mu < \mu^* \in M$ are satisfied. If the main diagonal minors of the matrix $S^*(\mu)$ are positive at all $\mu < \mu^*$, then the trivial solution of the system (5.3.2) is globally exponentially μ-stable.*

Proof Prove statement (1). For the function $v(x, \theta)$ in the form (5.5.2) it is not difficult to obtain the estimates (5.5.3) at all $x \in \Pi$. In addition,

$$Dv(x, \theta)|_{(5.3.2)} \leq -\frac{1}{2}u^{\mathrm{T}} \left(\theta S(\mu) + S^{\mathrm{T}}(\mu)\theta\right) u, \qquad (5.5.11)$$

where $u^{\mathrm{T}} = \left(\psi_{13}^{1/2}(\|x_1\|), \ldots, \psi_{m3}^{1/2}(\|x_m\|)\right)$, $S(\mu)$ is an $(m \times m)$-matrix with the elements (5.5.10), and $\theta = \mathrm{diag}\,[\theta_1, \ldots, \theta_m]$. It is known that the conditions for the positiveness of the main diagonal minors of the matrix $S(\mu)$ are equivalent to the existence of a diagonal matrix θ with positive elements, such that the matrix $\left(\theta S(\mu) + S^{\mathrm{T}}(\mu)\theta\right)$ is positive definite at all $\mu < \mu^* \in M$. In this case $\theta_i = 1$, $i = 1, 2, \ldots, m$, and this condition is satisfied. Thus, the estimate (5.5.11) takes the form

$$Dv(x, \theta)|_{(5.3.2)} \leq \lambda_M(S)\psi_3(\|x\|) \qquad (5.5.12)$$

at all $x \in \Pi$ and $\lambda_M(S) < 0$ at $\mu < \mu^*$. The estimate (5.5.12) and Theorem 5.4.2 imply statement (1) of Theorem 5.5.5.

Statements (2)–(4) of this theorem are proved in a similar manner.

Now we turn our attention to the conditions for the μ-instability of the trivial solution of the system (5.3.2).

Assumption 5.5.6 An isolated subsystem from the collection (5.3.1) permits the property D, if there exists a semigroup $Q_i(t)$ and a function $v_i(x_i)$ which form a permissible pair $(Q_i(t), v_i)$, comparison functions $\psi_{i1}, \psi_{i2}, \psi_{i3}$ from K-class, and real constants β_i and Δ_i such that

(a) $\psi_{i1}(\|x_i\|) \leq v_i(x_i) \leq \psi_{i2}(\|x_i\|)$,

(b) $\nabla v_i(x_i, {}^s f_i(x_i)) \geq \beta_i \psi_{i3}(\|x_i\|)$ at all $x_i \in \mathcal{D}({}^s f_i)$, where $\mathcal{D}({}^s f_i)$ denotes the domain of definition of the infinitesimal generator of the semigroup $Q_i(t)$ at $\|x_i\| < \Delta_i$.

Assumption 5.5.7 The operator of connection between the subsystems (5.3.1) satisfies the property E, if at a specified permissible pair $(Q_i(t), v_i)$ there exist comparison functions ψ_{i3} from K-class and constants $c_{ij}(\mu)$, $i, j = 1, 2, \ldots, m$, such that

$$\nabla v_i(x_i, g_i(x, \mu)) \geq \psi_{i3}^{1/2}(\|x_i\|) \sum_{j=1}^{m} c_{ij}(\mu) \psi_{i3}^{1/2}(\|x_i\|)$$

at all $x^T = (x_1, \ldots, x_m) \in \mathcal{D}(f + g(x, \mu))$ and $\|x_i\| < \Delta_i$, $i = 1, 2, \ldots, m$.

Note that if in condition (b) of Assumption 5.5.6 the quantities $\beta_i > 0$, $i = 1, 2, \ldots, m$, then the trivial solution of all independent subsystems (5.3.1) is unstable.

Consider the following statement.

Theorem 5.5.6 *Assume that for each subsystem of the system (5.3.2) there exists a semigroup $Q_i(t)$ and a function $v_i(x_i)$ which form a permissible pair $(Q_i(t), v_i)$, and, in addition:*

(1) *isolated subsystems from the collection (5.3.1) permit property D;*

(2) *the operators of connection $g_i(x, \mu)$ between the subsystems of (5.3.1) permit property E;*

(3) *there exist constants $\theta_i > 0$, $i = 1, 2, \ldots, m$, and a value of the parameter $\mu^* \in M$ such that the matrix $C(\mu)$ with the elements*

$$c_{ij}(\mu) = \begin{cases} \theta_i(\beta_i + c_{ii}(\mu)) & at \quad i = j, \\ \frac{1}{2}(\theta_i c_{ij}(\mu) + \theta_j c_{ji}(\mu)) & at \quad i \neq j \end{cases}$$

is positive definite at all $\mu < \mu^*$.

Then the trivial solution of the system (5.3.2) is μ-unstable.

Proof For the function (5.5.2) under the conditions of Theorem 5.5.6 it is easy to obtain the estimates

$$\psi_1(\|x\|) \leq v(x, \theta) \leq \psi_2(\|x\|) \tag{5.5.13}$$

at all $x \in \Pi$ and

$$Dv(x,\theta)\big|_{(5.3.2)} \geq \lambda_m(C)\psi_3(\|x\|) \tag{5.5.14}$$

at all $x \in \Pi$, where $\lambda_m(C) > 0$ is the minimum eigenvalue of the matrix $C(\mu)$ at $\mu < \mu^*$. The estimates (5.5.13) and (5.5.14) and Theorem 5.4.6 imply that the trivial solution of the system (5.3.2) is μ-unstable.

Let us turn now to the properties of the μ-boundedness of the motion of the system (5.3.2).

Assumption 5.5.8 An isolated subsystem from the collection (5.3.1) permits the property F, if there exist a semigroup $Q_i(t)$ and a function $v_i(x_i)$ which form a permissible pair $(Q_i(t), v_i)$, comparison functions $\psi_{i1}, \psi_{i2}, \psi_{i3}$ from KR-class, and real constants β_i^*, such that:

(1) $\psi_{i1}(\|x_i\|) \leq v_i(x_i) \leq \psi_{i2}(\|x_i\|)$,

(2) $\nabla v_i(x_i, {}^sf_i(x_i)) \leq \beta_i^* \psi_{i3}(\|x_i\|)$ at all $x_i \in \mathcal{D}({}^sf_i)$ and

    (a) at all $\|x_i\| > \Delta_i^*$,

    (b) if $|v_i(x_i)| \leq m_i$, $|\nabla v_i(x_i, {}^sf_i(x_i))| \leq m_i$ at $\|x_i\| \leq \Delta_i^*$, where $m_i > 0$ is const.

Assumption 5.5.9 The operators of connection $g_i(x,\mu)$ between the subsystems of (5.3.1) satisfy property G, if at a specified permissible pair $(Q_i(t), v_i)$ there exist real constants $b_{ij}(\mu)$, $i,j = 1,2,\ldots,m$, such that

$$\nabla v_i(x_i, g_i(x,\mu)) \leq \psi_{i3}^{1/2}(\|x_i\|) \sum_{j=1}^{m} b_{ij}(\mu)\psi_{j3}^{1/2}(\|x_i\|)$$

at all $x^{\mathrm{T}} = (x_1, \ldots, x_m) \in \mathcal{D}(f + g(x,\mu))$.

Consider the following statement.

Theorem 5.5.7 *Assume that for each subsystem of the system (5.3.2) there exists a semigroup $Q_i(t)$ and a function $v_i(x_i)$ which form a permissible pair, and, in addition:*

(1) *isolated subsystems from the collection (5.3.1) permit property F;*

(2) *the operators of connection $g_i(x,\mu)$ between the subsystems of (5.3.1) permit property G;*

(3) *there exist constants $\theta_i > 0$, $i = 1,2,\ldots,m$, and a value of the parameter $\mu^* \in M$ such that the matrix $B(\mu) = [b_{ij}(\mu)]$ with the elements*

$$b_{ij}(\mu) = \begin{cases} \theta_i(\beta_i^* + b_{ii}(\mu)) & at \quad i = j, \\ \frac{1}{2}(\theta_i b_{ij}(\mu) + b_{ji}(\mu)\theta_j) & at \quad i \neq j \end{cases}$$

is negative definite at all $\mu < \mu^*$.

Then the motion of the system (5.3.2) is uniformly ultimately μ-bounded.

Proof Consider the function (5.5.2). Under the conditions of Assumptions 5.5.8 and 5.5.9 for the functions $v(x,\theta)$ and $Dv(x,\theta)$, obtain the estimates

$$\psi_1(\|x\|) \le v(x,\theta) \le \psi_2(\|x\|) \tag{5.5.15}$$

and

$$Dv(x,\theta)|_{(5.3.2)} \le \lambda_M(B)\psi_3(\|x\|) \tag{5.5.16}$$

at all $x \in \mathbb{X} - \prod\limits_{i=1}^{m} \overline{S}_i(m_i)$, where $\overline{S}_i(m_i) = \{x_i \in \mathbb{Z}_i : \|x_i\| \le m_i\}$.

Consider the estimates (5.5.15) and (5.5.16) in two cases.

Case 1. Let $x_i \in \mathbb{Z}_i$ and $\|x_i\| > m_i$ at $i = 1, 2, \ldots, p$.

Case 2. For the values $i = p+1, \ldots, m$ $x_i \in \mathbb{Z}_i$ and $\|x_i\| \le m_i$ at $x^{\mathrm{T}} = (x_1, \ldots, x_m) \in \mathcal{D}(f + g(x,\mu))$.

The estimates (5.5.15) are transformable to the following:

$$\sum_{i=1}^{p} \theta_i \psi_{i1}(\|x_i\|) + \sum_{i=p+1}^{m} \theta_i v_i(x_i) \le v(x,\theta) \le \sum_{i=1}^{p} \theta_i \psi_{i2}(\|x_i\|) + \sum_{i=p+1}^{m} \theta_i v_i(x_i)$$

in Case 1 and

$$\sum_{i=1}^{p} \theta_i \psi_{i1}(\|x_i\|) - \sum_{i=p+1}^{m} \theta_i m_i \le v(x,\theta) \le \sum_{i=1}^{p} \theta_i \psi_{i2}(\|x_i\|) + \sum_{i=p+1}^{m} \theta_i m_i$$

in Case 2.

For the expression $\nabla v(x, f(x) + g(x,\mu))$ obtain the estimate

$$\nabla v(x, f(x) + g(x,\mu)) \le w^{\mathrm{T}} B^*(\mu) w + \sum_{i=1}^{p} \theta_i \psi_{i3}^{1/2}(\|x_i\|) \left[\sum_{j=p+1}^{m} b_{ij}(\mu) \psi_{j3}^{1/2}(m_j) \right]$$

$$+ \sum_{i=1}^{p} \theta_i m_i + \sum_{i=p+1}^{m} \theta_i \psi_{i3}^{1/2}(m_i) \sum_{j=1}^{p} b_{ij}(\mu) \psi_{j3}^{1/2}(\|x_i\|)$$

$$+ \sum_{i=p+1}^{m} \theta_i \psi_{i3}^{1/2}(m_i) \sum_{j=p+1}^{m} \psi_{i3}^{1/2}(m_j), \tag{5.5.17}$$

where $B^*(\mu) = [b_{ij}(\mu)]$ at $i, j = 1, 2, \ldots, p$ and $w = (\psi_{13}^{1/2}(\|x_1\|), \psi_{23}^{1/2}(\|x_2\|), \ldots, \psi_{p3}^{1/2}(\|x_p\|))^{\mathrm{T}}$.

It is easy to reduce the estimate (5.5.17) to the form

$$\nabla v(x, f(x) + g(x,\mu)) \le w^{\mathrm{T}} B^*(\mu) w + w^{\mathrm{T}} P_0 + P_1, \tag{5.5.18}$$

where $P_0 \in \mathbb{R}^p$ and $P_1 > 0$ is some constant.

Now, since the matrix $B(\mu)$ is negative definite, the submatrix $B^*(\mu)$ will also be negative definite at $\mu < \mu^*$. Then the estimate (5.5.18) will take the form

$$\nabla v(x, f(x) + g(x, \mu)) \leq \lambda_M(B^*)\|w\|^2 + w^{\mathrm{T}} P_0 + P_1$$

$$\leq \frac{1}{2}\lambda_M(B^*)\|w\|^2 = \frac{1}{2}\lambda_M(B^*) \sum_{i=1}^{p} \psi_{i3}(\|x_i\|)$$

$$\leq \frac{1}{2}\lambda_M(B^*)\psi_3^*\left(\sum_{i=1}^{p} \|x_i\|\right),$$

(5.5.19)

where ψ_3^* belongs to the KR-class. Since $\lambda_M(B^*) < 0$, the estimate (5.5.19) implies that $Dv(x, \theta) < 0$ at all $x^{\mathrm{T}} = (x_1, \ldots, x_m) \in \mathcal{D}(f + g(x, \mu))$, $0 < \mu < \mu^*$ and at $\|x_i\| > r^*$ for $i = 1, 2, \ldots, p$, and $\|x_i\| \leq m_i$ for $i = p + 1, \ldots, m$. According to Theorem 5.4.8, the motion of the system (5.3.2) is uniformly ultimately μ-bounded.

5.6 Stability Analysis of a Two-Component System

Consider a physical process described by the system of equations

$$\frac{dx_1}{dt} = f_1(x_1(t)) + \mu b \int_G H_1(y, x_2(t, y))\, dy,$$

$$\frac{\partial x_2(t, y)}{\partial t} = \alpha \Delta x_2(t, y) - H_2(x_2(t, y)) + \mu h_2(y) c^{\mathrm{T}} x_1(t)$$

(5.6.1)

with the boundary

$$x_2(t, y) = 0 \quad \text{at all} \quad (t, y) \in \mathbb{R}_+ \times \partial G$$

(5.6.2)

and the initial

$$x_1(0) = x_{10}, \quad x_2(0, y) = \psi(y) \quad \text{at} \quad y \in G$$

(5.6.3)

conditions. Here $f(x) \colon \mathbb{R}^n \to \mathbb{R}^n$, b, c are specified n-dimensional vectors, α and L are specified positive constants, Δ is a Laplace operator in the space $\mathbb{R}^m$, G is an open subset in $\mathbb{R}^m$ with a smooth boundary ∂G, and $\mu \in M$ is a small positive parameter. The functions H_1 and H_2 are specified and satisfy the conditions

(a) $H_1(y, 0) = 0$ at all $y \in G$,

(b) $H_2(0) = 0$ and $|H_1(y, z) - H_1(y, z^*)| \leq |h_1(y)|\|z - z^*\|$ at all $y \in G$, and

(c) $z, z^* \in \mathbb{R}$ and $|H_2(u) - H_2(u^*)| \leq L\|u - u^*\|$ at all $u, u^* \in \mathbb{R}$ and $h_i \in L_2(G)$, $i = 1, 2$.

Under the above conditions the problem (5.6.1) – (5.6.3) is correctly defined and its solution $(x_1(t), x_2(t, y))^\mathrm{T}$ exists at all $t \in \mathbb{R}_+$.

The isolated subsystems of the system (5.6.1) have the form

$$\frac{dx_1}{dt} = f_1(x_1), \tag{5.6.4}$$

$$\frac{\partial x_2(t, y)}{\partial t} = \alpha \Delta x_2(t, y) - H_2(y) \triangleq f_2(x_2). \tag{5.6.5}$$

The operators of connection $g_i(x, \mu)$, $i = 1, 2$, are as follows:

$$\mu g_1(x_1, x_2) = \mu b \int_G H_1(y, x_2(t, y))\, dy, \tag{5.6.6}$$

$$\mu g_2(x_1, x_2) = \mu h_2(y) c^\mathrm{T} x_1(t). \tag{5.6.7}$$

For the system (5.6.1) it is assumed that $\mathbb{Z}_1 = \mathbb{R}^n$, $\mathbb{Z}_2 = L_2(G)$, and $\mathbb{X} = \mathbb{R}^n \times L_2(G)$. The norms in $\mathbb{R}^n$ and on $L_2(G)$ will be denoted by $\|\cdot\|$ and $\|\cdot\|_{L_2}$, respectively.

Assumption 5.6.1 There exist:

(1) functions $v_{11}(x_1) \in C(\mathbb{R}^n, \mathbb{R}_+)$ and $v_{22}(x_2) \in C(L_2(G), \mathbb{R}_+)$ in open connected neighborhoods of the points $x_1 = 0$, $x_2 = 0$, comparison functions $\varphi_i(\|x_1\|)$ and $\psi_i(\|x_2\|_{L_2})$ from K-class, and positive constants $\underline{\alpha}_{ii}$, $\overline{\alpha}_{ii}$, $i = 1, 2$, such that

$$\underline{\alpha}_{11}\varphi_1^2(\|x_1\|) \le v_{11}(x_1) \le \overline{\alpha}_{11}\varphi_2^2(\|x_1\|),$$
$$\underline{\alpha}_{22}\psi_1^2(\|x_2\|_{L_2}) \le v_{22}(x_2) \le \overline{\alpha}_{22}\psi_2^2(\|x_2\|_{L_2});$$

(2) functions $v_{12}(x_1, x_2) = v_{21}(x_1, x_2) \in C(\mathbb{R}^n \times L_2(G), \mathbb{R})$ and arbitrary constants $\underline{\alpha}_{12}$, $\overline{\alpha}_{21}$ such that

$$\underline{\alpha}_{12}\varphi_1(\|x_1\|)\psi_2(\|x_2\|_{L_2}) \le v_{12}(x_1, x_2) \le \overline{\alpha}_{12}\varphi_2(\|x_1\|)\psi_2(\|x_2\|_{L_2})$$

in the range of values $x_1 \in \mathcal{D}(f_1)$ and $x_2 \in \mathcal{D}(f_2)$.

Lemma 5.6.1 *If all the conditions of Assumption 5.6.1 are satisfied and the matrices*

$$A_1 = \begin{pmatrix} \underline{\alpha}_{11} & \underline{\alpha}_{12} \\ \underline{\alpha}_{21} & \underline{\alpha}_{22} \end{pmatrix}, \quad \underline{\alpha}_{12} = \underline{\alpha}_{21},$$

$$A_2 = \begin{pmatrix} \overline{\alpha}_{11} & \overline{\alpha}_{12} \\ \overline{\alpha}_{21} & \overline{\alpha}_{22} \end{pmatrix}, \quad \overline{\alpha}_{12} = \overline{\alpha}_{21},$$

are positive definite, then the function $v(x, \theta) = \theta^\mathrm{T} U(x)\theta$, $\theta \in \mathbb{R}_+^2$, $U(x) = [u_{ij}(\cdot)]$, $i, j = 1, 2$, is positive definite and decreasing.

Assumption 5.6.2 For the specified functions $v_{11}(x_1)$, $v_{22}(x_2)$, and $v_{12}(x_1, x_2)$ there exist constants β_{ik}, $i = 1, 2$, $k = 1, 2, \ldots, 8$, comparison functions $\xi_1(\|x_1\|)$ and $\xi_2(\|x_2\|_{L_2})$ from K-class, such that:

(a) $\nabla v_{11}(x_1, f_1(x_1)) \leq 0$;

(b) $\nabla v_{11}(x_1, g_1(x, \mu)) \leq \beta_{12}\xi_1^2(\|x_1\|) + \beta_{13}\xi_1(\|x_1\|)\xi_2(\|x_2\|_{L_2})$;

(c) $\nabla v_{22}(x_2, f_2(x_2)) \leq 0$;

(d) $\nabla v_{22}(x_2, g_2(x, \mu)) \leq \beta_{22}\xi_2^2(\|x_2\|_{L_2}) + \beta_{23}\xi_1(\|x_1\|)\xi_2(\|x_2\|_{L_2})$;

(e) $\nabla v_{12}(x_1, x_2, f_1(x_1)) \leq \beta_{14}\xi_1^2(\|x_1\|) + \beta_{15}\xi_1(\|x_1\|)\xi_2(\|x_2\|_{L_2})$;

(f) $\nabla v_{12}(x_1, x_2, f_2(x_2)) \leq \beta_{24}\xi_1^2(\|x_1\|) + \beta_{25}\xi_1(\|x_1\|)\xi_2(\|x_2\|_{L_2})$;

(g) $\nabla v_{12}(x_1, x_2, g_1(x, \mu)) \leq \beta_{16}\xi_1^2(\|x_1\|) + \beta_{17}\xi_1(\|x_1\|)\xi_2(\|x_2\|_{L_2}) + \beta_{18}\xi_2^2(\|x_2\|_{L_2})$;

(h) $\nabla v_{12}(x_1, x_2, g_2(x, \mu)) \leq \beta_{26}\xi_1^2(\|x_1\|) + \beta_{27}\xi_1(\|x_1\|)\xi_2(\|x_2\|_{L_2}) + \beta_{28}\xi_2^2(\|x_2\|_{L_2})$.

Consider the matrix $C(\mu)$ in the form

$$C(\mu) = \begin{pmatrix} c_{11} & c_{12} \\ c_{21} & c_{22} \end{pmatrix}, \quad c_{12} = c_{21},$$

with the elements

$$c_{11} = \theta_1^2\mu\beta_{12} + 2\theta_1\theta_2(\beta_{14} + \mu\beta_{16} + \mu\beta_{26}),$$

$$c_{22} = \theta_2^2\mu\beta_{22} + 2\theta_1\theta_2(\beta_{24} + \mu\beta_{18} + \mu\beta_{28}),$$

$$c_{12} = \frac{1}{2}\left(\theta_1^2\mu\beta_{13} + \theta_2^2\mu\beta_{23}\right) + \theta_1\theta_2(\beta_{15} + \beta_{25} + \mu\beta_{17} + \mu\beta_{27}).$$

Introduce the notation a, p, q, μ_1, μ_2, μ_3, μ_4 by the formulae

$$a = \theta_1\theta_2[\theta_1\beta_{12} + 2\theta_1(\beta_{16} + \beta_{26})][\theta_2\beta_{22} + 2\theta_1(\beta_{13} + \beta_{28})]$$
$$- \left[\frac{1}{2}(\theta_1^2\beta_{13} + \theta_2^2\beta_{23}) + \theta_1\theta_2(\beta_{17} + \beta_{27})\right]^2,$$

$$p = \theta_1\theta_2\Big\{[\theta_1\beta_{24}(\theta_1\beta_{12} + 2\theta_1\beta_{16} + 2\theta_2\beta_{26})$$
$$+ \theta_2\beta_{14}(\theta_2\beta_{22} + 2\theta_1\beta_{18} + 2\theta_1\beta_{28})]$$
$$- (\beta_{15} + \beta_{25})\left[\frac{1}{2}(\theta_1^2\beta_{13} + \theta_2^2\beta_{23}) + \theta_1\theta_2(\beta_{17} + \beta_{27})\right]\Big\},$$

$$q = \theta_1^2\theta_2^2[4\beta_{14}\beta_{24} - (\beta_{15} + \beta_{25})^2],$$

$$\mu_1 = -2\frac{\theta_1}{\theta_2}\frac{\beta_{14}}{\beta_{12}} + 2\theta_2(\beta_{16} + \beta_{26}),$$

$$\mu_2 = -2\frac{\theta_1}{\theta_2}\frac{\beta_{24}}{\beta_{22}} + 2\theta_1(\beta_{18} + \beta_{28}),$$

$$\mu_3 = (p + \sqrt{p^2 - 4aq})(-2a)^{-1},$$

$$\mu_4 = (-p - \sqrt{p^2 - 4aq})(-2a)^{-1}.$$

Now consider the value of the parameter μ for which the boundary value μ_0 is determined from the conditions:

B_0. If $a > 0$, $p > 0$, then $\mu_0 = \min(\mu_1, \mu_2)$.

B_1. If $a < 0$, p is arbitrary, then $\mu_0 = \min(\mu_1, \mu_2, \mu_3)$.

B_2. If $a > 0$, $p < 0$, then $\mu_0 = \min(\mu_1, \mu_2, \mu_4)$.

It is not difficult to verify the correctness of the following lemma.

Lemma 5.6.2 *If all the conditions of Assumption 5.6.2 are satisfied and the inequalities*

(a) $\theta_1 \beta_{14} < 0$;

(b) $\theta_1 \beta_{12} + 2\theta_2(\beta_{16} + \beta_{26}) > 0$;

(c) $\theta_1 \beta_{24} < 0$;

(d) $\theta_2 \beta_{22} + 2\theta_1(\beta_{18} + \beta_{28}) > 0$;

(e) $4\beta_{14} - (\beta_{15} + \beta_{25})^2 > 0$

hold, then the matrix $C(\mu)$ is negative definite at $\mu \in (0, \mu_0)$, where μ_0 is determined by one of the conditions $B_0 - B_2$.

Taking into account the condition of Assumption 5.6.2, for the function $v(x, \theta)$ we obtain the following estimate of the derivative:

$$\nabla v(x, \theta)\big|_{(5.6.1)} \leq u^T C(\mu) u, \qquad (5.6.8)$$

where $u^T = (\xi_1(\|x_1\|), \xi_2(\|x_2\|))$, $\mu \in (0, \mu_0)$.

Theorem 5.6.1 *If the two-component system (5.6.1) and (5.6.2) is such that all the conditions of Lemmas 5.6.1 and 5.6.2 are satisfied, then its state of equilibrium $x_1 = 0$, $x_2 = 0$ is uniformly asymptotically μ-stable at $\mu \in (0, \mu_0)$.*

The proof of the theorem follows from the conditions satisfied by the function $v(x, \theta)$ and its derivative (5.6.8).

Remark 5.6.1 In view of conditions (a) and (c) from Assumption 5.6.2, the hybrid system (5.6.1) and (5.6.2) consists of stable (nonasymptotically) subsystems, and the uniform asymptotic μ-stability of the state of equilibrium $x_1 = 0$, $x_2 = 0$ is achieved due to the stabilizing influence of the connection operators.

5.7   Comments and References

Hybrid dynamical systems, being systems consisting of two or more different subsystems connected with each other, are widely spread models of real processes and phenomena (see Haddad, Chellaboina, and Nersesov [1] and bibliography therein). Initially the class of hybrid systems included those whose dynamics were described by systems of ordinary differential equations on $\mathbb{R}_+$ and systems of difference equations on $\mathbb{Z}$. Examples of such systems are systems with impulse effects (see Bainov and Simeonov [1], Samoilenko and Perestyuk [1], and others), switched systems (see Branicky [1], Peleties and De Carlo [1]), systems with variable structure (see Utkin [1]), and other systems. The concept of generalized time (see Michel [1]) made it possible to unify a lot of results obtained in this line of investigation by way of consideration of a generalized hybrid system in a metric space (see Michel, Wang, and Hu [1], Martynyuk [18]).

A more general class is the class of hybrid systems consisting of different subsystems connected by operators (see Matrosov [1], Matrosov and Vassiliev [1], and others). In this chapter, in compliance with the concept of this book, hybrid systems with weakly connected subsystems described by equations in a Banach space are considered.

5.2. The proofs of the statements given in this section are available in the monographs of Hille and Phillips [1] and Krein [1]; see also Crandall [1], Brezis [1], and Kurtz [1].

5.3. In the statement of the problem on μ-stability of solutions of a hybrid system, the results of the articles of Martynyuk [8, 10] were considered. Note that a similar problem was considered in the monograph of Michel and Miller [1] and others.

5.4. Originally, a two-index system of functions as an environment suitable for construction of a Lyapunov function was considered in the works of Djordjević [1], Martynyuk and Gutovsky [1], and Martynyuk [7, 9] for systems of ordinary differential equations. For equations in a Banach space, a matrix function was applied in the work of Martynyuk [8]. Theorems 5.4.1–5.4.8 are analogues of the classical theorems of the general theory of stability and new for this class of hybrid systems. In the works of Lakshmikantham [1, 2], Massera [1], and Zubov [1], one can find some approaches to the analysis of the stability of solutions of equations in Banach spaces, which may be generalized for hybrid systems.

5.5. All the results of this section are new for the class of systems of (5.3.2) type. Theorems 5.5.3 and 5.5.4 on the exponential μ-stability of a hybrid system are formulated and proved in view of the results of the works of He and Wang [1] and Martynyuk [14, 15].

5.6. The results of this section are new for the system (5.6.1). Some of the results were taken from the work of Martynyuk [9]. A system of the type (5.6.1) at $f(x_1) = Ax_1$ and $\mu = 1$ was studied in the monograph by Michel and Miller [1] on the basis of a vector Lyapunov function. The assumption on the asymptotic stability of the zero solution of the independent subsystems (5.6.4) and (5.6.5) makes it possible to apply the vector Lyapunov function, but the connection operators $g_i(x)$, $i = 1, 2$ are considered as factors that destabilize the trivial solution of the system under consideration.

Bibliography

Aizerman M.A., Gantmacher F.R.

[1]   *Absolute stability of controllable systems.* Moscow: Publ. of USSR Acad. Sci., 1963.

Aminov A.B., Sirazetdinov T.K.

[1]   The method of Lyapunov functions in the problem of polystability of motion. *Appl. Math. Mech.* **51** (1987) 553–558.

Anashkin O.V.

[1]   On asymptotic stability in nonlinear systems. *Differential Equations,* **14** (8) (1978) 1490–1493.

[2]   *Asymptotic methods in the stability theory.* Simferopol: SGU, 1983.

Bainov D.D., Simeonov P.S.

[1]   *Systems with impulse effects: stability, theory and applications.* New York: Halsted, 1989.

Barbashin E.A.

[1]   *Lyapunov functions.* Moscow: Nauka, 1970.

Barbashin E.A., Krasovsky N.N.

[1]   On global stability of motion. *Dokl. Acad. Nauk SSSR,* **86** (3) (1952) 453–456.

Beesack P.R.

[1]   *Gronwal inequalities.* Cavleton Math. Lecture Notes **11**, 1975.

Bellman R.

[1]   *Theory of stability of motions of differential equations.* Moscow: Publishing House of Foreign Literature, 1954.

Bihari I.

[1]   A generalization of a lemma of Bellman and its application to uniqueness problems of differential equations. *Acta Math. Acad. Sci. Hung.* **7** (1) (1956) 71–94.

Bogolyubov N.N., Mitropolsky Yu. A.

[1]   *Asymptotic methods in the theory of nonlinear oscillations.* Moscow: Nauka, 1974.

Bourland F.J., Haberman R.

[1]   The modulated phase shift for strongly nonlinear slowly varying and weakly damped oscillators. *SIAM J. Appl. Math.* **48** (1988) 737–748.

Branicky M.S.

[1]   Stability of switched and hybrid systems. *Proc. of the 33rd Conf. on Decision and Control.* (1994) 3498—3508.

Brezis H.

[1]   *Operateurs maximaux monotones.* Amsterdam: North-Holland Publ., 1973.

Burdina V.I.

[1]   On the boundedness of solutions of differential equations. *Dokl. Acad. Nauk SSSR,* **93** (1953) 603–606.

Burton T.A.

[1]   Liapunov functions and boundedness. *J. Math. Anal. Appl.* **58** (1) (1977) 88–97.

[2]   The generalized Lienard equation. *SIAM J. Control* **3** (1965) 223–230.

Česari L.

[1]   *Asymptotic behaviour and stability of solutions of ordinary differential equations.* Moscow: Mir, 1964.

Chernetskaya L.N.

[1]   *Generalized entry and estimates of stability of motions of nonautonomous large mass systems.* PhD thesis, Inst. of Mechanics of National Acad. of Sci. Ukr., 1986.

[2]   On stability on a finite interval. *Appl. Mech.,* **23**(7) (1987) 92–97.

Chetaev N.G.

[1]   *Stability of motion.* Moscow: Nauka, 1990.

Coddington E.A., Levinson N.

[1]   *Theory of ordinary differential equations.* New York: McGill-Hill, 1955.

Conti R.

[1]   Sulla prolungabilità delle soluzioni di in sistema di equazioni differenziali or-dinarie. *Doll. Un. Mat. Ital.,* **11**(3) (1956) 510–514.

Corduneanu C.

[1]   On the existence of bounded solutions for some nonlinear systems. *Dokl. Acad. Nauk SSSR,* **131** (4) (1960) 735–737.

[2]   Application of didfferential inequalities in the stability theory. *An. Sti. Univ. "Al. I. Cuza", Iasi , Sect. I a Mat.* **6** (1960) 47–58.

[3]   Sur la stabilite partielle. *Rev. Roumanie Math. Pure Appl.,* **9** (1964) 229–236.

Crandall M.G.

[1]   Semigroups of nonlinear transformations in Banach spaces. In: *Contribution to nonlinear functional analysis.* (Ed.: E.H. Zarantonello) New York: Academic Press, 1971.

Demidovich B.P.

[1]   On bounded solutions of some nonlinear system of ordinary differential equations. *Math. Coll.,* **40 (82)** (1) (1956) 73–94.

Djordjevič, M.Z.

[1]   Stability analysis of interconnected systems with possible unstable subsystems. *Systems and Control Letters* **3** (1983) 165–169.

Duboshin G.N.

[1]   On the question of stability of motion with respect to continuous perturbations *Proc. of State Astron. Inst. named by Sternberg,* **14** (1940) 153–164.

[2]   A problem of stability under continuous perturbations. *Vestn. Mosc. Univ.,* (2) (1952) 35–40.

Euler L.
[1]   *The method of finding curve lines having a property of maximum or minimum.* Moscow: GTTI, 1934.

Goisa L.N., Martynyk A.A.
[1]   Study of a systems of oscillators with small interaction in a neighborhood of a singular point. *Math. Phys.*, **15** (1974) 34–43.

Gorshin S.I.
[1]   On some criteria of stability under continuous perturbations. *Proc. of Higher Education Establishments, Mathematics*, (11) (1967) 17–20.

Grebenikov E.A.
[1]   *Averaging method in applied problems.* Moscow: Nauka, 1986.

Grujić Lj. T., Martynyuk A.A., Ribbens-Pavella M.
[1]   *Large-scale systems stability under structural and singular perturbations.* Kiev: Naukova Dumka, 1984.

Gusarova R.S.
[1]   On boundedness of solutions of a linear differential equation with periodic coefficients. *Appl. Math. Mech.*, **13** (3) (1949) 241–246.

Haddad W.M., Chellaboina V.S., Nersesov S.G.
[1]   *Impulsive and hybrid dynamical systems.* Princeton and Oxford: Princeton Univ. Press, 2006.

Hahn W.
[1]   *Stability of motion.* Berlin: Springer, 1967.

Hatvani L.
[1]   On partial asymptotic stability and instability. I (Autonomous systems). *Acta Sci. Math.*, **45** (1983) 219–231.

Hayashi T.
[1]   *Nonlinear oscillations in physical systems.* Moscow: Mir, 1968.

He J., Wang M.
[1]   Remarks on exponential stability by comparison functions of the same order of amplitude. *Ann. Diff. Eqns.* **7**(4) (1991) 409–414.

Hille E., Phillips R.S.
[1]   *Functional analysis and semi-groups.* Providence, RI: Amer. Mat. Soc., 1957.

Hoppensteadt F.C.
[1]   *Analysis and simulation of chaotic systems.* New York: Springer, 1993.

Izhikevich E.M., Hoppensteadt F.C.
[1]   *Weakly connected neural networks.* New York: Springer, 1997.

Kamenkov G.V.
[1]   *Stability of motion. Oscillations. Aerodynamics.* Moscow: Nauka, 1972.

Karimzhanov K.
[1]   *Development of the second Lyapunovs method in the study of stability of unsteady motions on the basis of auxiliary systems.* PhD thesis, Inst. of Mechanics of National Acad. of Sci. of Ukraine, 1985.

Karimzhanov K., Kosolapov V.I.
[1]   On the use of limit systems in the study of stability with the aid of Lyapunov functions. *Math. Phys. Nonlin. Mechanics*, **1** (1984) 32–37.

Khapayev M.M.

[1]   *Averaging in the stability theory.* Moscow: Nauka, 1986.

Kosolapov V.I.

[1]   On the stability of motion in a neutral case. *Dokl. Acad. Nauk Ukr. SSR, Ser. A.*, (1) (1979) 27–31.

[2]   On the stability of complex systems. *Appl. Mech.* **15** (7) (1979) 133–136.

[3]   *The second Lyapunovs method in the problem of stability of nonlinear systems with integrable approximation.* PhD thesis, Inst. of Mechanics of National Acad. of Sci. of Ukraine, 1979.

[4]   Investigation of asymptotic stability and instability of nonlinear systems with integrable approximation. *Math. Phys.*, **29** (1981) 44–48.

Krasovsky N.N.

[1]   *Some problems of the theory of stability of motion.* Moscow: GIFML, 1959.

Krein S.G.

[1]   *Linear differential equations in Banach space.* Providence, RI: Amer. Mat. Soc., 1970.

Krylov N.M., Bogolyubov N.N.

[1]   *Investigation of longitudinal stability of an aeroplane.* Kiev: Publ. of Ukr. SSR Acad. of Sci., 1932.

[2]   *Introduction into nonlinear mechanics.* Kiev: Publ. of Ukr. Acad. of Sci., 1937.

Kurtz T.

[1]   Convergence of sequences of semigroups of nonlinear equations with applications to gas kinetics. *Trans. Amer. Mat. Soc.* **186** (1973) 259–272.

Lagrange G.

[1]   *Analytical mechanics.* V.1. Moscow–Leningrad: Gostechizdat, 1938.

Lakshmikantham V.

[1]   Stability and asymptotic behavior of solutions of differential equations in a Banach space. *Lecture Notes*, CIME, Italy, 1974, 39–98.

[2]   Differential equations in Banach spaces and extension of Lyapunov's method. *Proc. Camb. Phi. Soc.* **59** (1063) 343–381.

Lakshmikantham V., Leela S.

[1]   On perturbing Lyapunov functions. *Math. System Theory* **10** (1976) 85–90.

Lakshmikantham V., Leela S., Martynyuk A.A.

[1]   *Stability of motion: comparison method.* Kiev: Naukova Dumka, 1991.

Lakshmikantham V., Salvadori L.

[1]   On Massera type converse theorem in terms of two different measures. *Bullettino U.M.I.* **13-A** (5) (1976) 293–301.

Lakshmikantham V., Xinzhi Liu

[1]   *Stability analysis in terms of two measures.* Singapore: World Scientific, 1993.

Lefshets S.

[1]   *Stability of nonlinear automatic control systems.* Moscow: Mir, 1967.

Letov A.M.

[1]   *Stability of nonlinear controllable systems.* Moscow: GIFML, 1962.

Lurie A.I.
[1] *Some nonlinear problems of the theory of automatic control.* Moscow: Gostech-izdat, 1951.

Lyapunov A.M.
[1] *General problem of stability of motion.* Moscow-Leningrad: Gostechizdat, 1950.

Malkin I.G.
[1] On stability under continuous perturbations. *Appl. Math. Mech.* **8** (3) (1944) 241–245.
[2] *The theory of stability of motion.* Moscow: Gostechizdat, 1952.

Martynyuk A.A.
[1] Theorem of the type of Lyapunov theorem on stability of a multidimensional system. *Ukr. Math. J.*, **24** (4) (1972) 532–537.
[2] On instability of equilibrium of a multidimensional system consisting of "neutrally" unstable subsystems. *Appl. Mech.*, **8** (6) (1972) 77–82.
[3] Qualitative investigation of the behavior of weakly coupled oscillators in the neighborhood of the equilibrium. *Appl. Mech.* , **9** (7) (1973) 122–126.
[4] Averaging method in the theory of stability of motion. *Zagadn. Drgan. Nielin.*, **14** (1973) 71–79.
[5] On stability of motions in nonlinear mechanics. *Dokl. Akad. Nauk SSSR* **264** (5) (1982) 1073–1077.
[6] *Stability of motion of complex systems.* Kiev: Naukova Dumka, 1975.
[7] The Lyapunov matrix-function. *Nonlin. Anal.*, **8** (1984) 1234–1241.
[8] The Lyapunov matrix-function and stability of hybrid systems. *Appl. Mech.*, **21** (1985) 89–96.
[9] Lyapunov matrix-function and stability theory. *Proc. IMACS-IFAC Symp.* IDN, Villeneuve d'Ascq., France, 1986, 261–265.
[10] The Lyapunov Matrix function and stability of hybrid systems. *Nonlin. Anal.*, **10** (1986) 1449–1457.
[11] A theorem on polystability. *Dokl. Akad. Nauk SSSR* **318**(4) (1991) 808–818.
[12] Stability theorem for nonautonomous equations with small perturbations *Dokl. Akad. Nauk of Ukraine* (4) (1992) 7–10.
[13] A theorem on instability under small perturbations. *Dokl. Akad. Nauk of Ukraine* (6) (1992) 14–16.
[14] On exponential stability with respect to a part of variables. *Dokl. Akad. Nauk* **331** (1993) 17–19.
[15] Exponential polystability of separable motions. *Dokl. Akad. Nauk* **336** (1994) 446–447.
[16] *Stability analysis: nonlinear mechanics equations.* Amsterdam: Gordon and Breach Publishers, 1995.
[17] *Stability by Lyapunov's matrix function method with applications.* New York: Marcel Dekker, 1998.
[18] Stability of dynamical systems in metric spaces. *Nonlin. Dynamics and Systems Theory* **5**(2) (2005) 157–167.

Martynyuk A.A., Chernetskaya L.N.
[1] On the estimation of unsteady motions on time-dependent sets. *Appl. Mech.* **25**(3) (1989) 90–95.

Martynyuk A.A., Gutovsky R.
[1] *Integral inequalities and stability of motion.* Kiev: Naukova Dumka, 1979.

Martynyuk A.A., Karimzhanov A.
[1]   Limiting equations and stability of non-stationary motions. *J. Math. Anal. Appl.* **132** (1988) 101–108.
[2]   Stability of weakly connected large-scale systems. *Soviet Appl. Mech.* **25**(3) (1989) 129–133.

Martynyuk A.A., Kosolapov V.I.
[1]   Higher derivatives of Lyapunov functions and the stability of motion under continuous perturbations. *Math. Phys.*, (21) (1978) 37–43.
[2]   Comparison principle and averaging method in the problem on stability of nonasymptotically stable motions under continuous perturbations. Preprint 78.33. Institute of mathematics of Ukr. SSR, Acad. of Sci., 1978.

Martynyuk V.A.
[1]   On the polystability of dynamical systems. *J. Math. Anal. Appl.* **191** (1995) 466–472.
[2]   Stability of nonlinear weakly coupled systems. *Int. Appl. Mech.*, **31**(4) (2004) 312–316.
[3]   On the boundedness with respect to two measures of nonlinear systems with a small parameter (to appear).
[4]   On algebraic conditions of the boundedness of nonlinear weakly coupled systems (to appear).

Massera J.L.
[1]   Contribution to stability theory. *Annals of Math.* **64**(1) (1956) 182–206.

Matrosov V.M.
[1]   The method of vector Lyapunov functions in the analysis of complex systems with distributed parameters. *Automatics and Telemechanics* **1** (1973) 5–22.
[2]   *Method of Lyapunov vector functions: Analysis of dynamical properties of nonlinear systems.* Moscow: Fismatlit, 2001.

Matrosov V.M., Vassiliev S.N.
[1]   The comparison principle in the dynamics of controllable systems with distributed parameters. In: *Thes. of VII All-Union. meeting on control problems.* Minsk, 1977.

Michel A.N.
[1]   Recent trends in the stability analysis of hybrid dynamical systems. *IEEE Trans. on Circuits and Systems — 1: Fundamental Theory and Applications* **46** (1999) 120–134.

Michel A.N., Miller R.K.
[1]   *Qualitative analysis of large scale dynamical systems.* New York: Academic Press, 1977.

Michel A.N., Wang K., Hu B.
[1]   *Qualitative theory of dynamical systems. The role of stability preserving mappings.* New York: Marcel Dekker, 2001.

Mitropolsky Yu. A.
[1]   Slow processes in nonlinear systems with many degrees of freedom. *Appl. Math. Mech.* **14** (1950) 139–170.

Mitropolsky Yu. A., Martynyuk V.A.
[1]   On the boundedness of motions of nonlinear weakly coupled systems. *Appl. Mech.* **29** (3) (1993) 68–73.

Movchan A.A.
[1]   Stability of processes with respect to two metrics. *Appl. Math. Mech.*, **24** (6) (1960) 988 – 1001.

Nayfeh A.N., Mook D.T.
[1]   *Nonlinear oscillations.* New York: Wiley-Interscience, 1979.

Peiffer K., Rouche N.
[1]   Liapunov's second method applied to partial stability. *J. de Mecanique* **8**(2) (1969) 323 – 334.

Peleties P., De Carlo R.
[1]   Asymptotic stability of *m*-switched systems using Lyapunov-like functions. *Proc. of the 1991 American Control Conf.*, 1991, 1679–1684.

Persidsky K.P.
[1]   On the theory of stability of solutions of differential equations. *Uspechi Mat. Nauk*, **1**, issue 5–6 (new series) (1946) 250–255.

Pliss V.A.
[1]   *Nonlocal problems of the theory of oscillations.* Moscow: Nauka, 1964.

Poincare A.
[1]   *On curves determine by differential equations.* Moscow–Leningrad: GITTL, 1947.
[2]   *New methods of celestial mechanics.* Moscow: Nauka, 1971.

Rao M. Rama Mohana
[1]   *Ordinary differential equations. Theory and applications.* New Delhi-Madras: Affiliated East-West Press Pvt Ltd, 1980.

Reissig R., Sansone G., Conti R.
[1]   *Qualitative Theorie nichtlinearer Differentialgleichungen.* Roma: Edizioni Cremonese, 1963.

Rouche N., Habets P., Laloy M.
[1]   *Direct Lyapunov method in the stability theory.* Moscow: Mir, 1968.

Rozo M.
[1]   *Nonlinear oscillations and stability theory.* Moscow: Nauka, 1971.

Rumiantsev V.V.
[1]   On stability of motion with respect to a part of variables. *Vestn. Mosc. Univ.* **4** (1957) 9 – 16.

Rumiantsev V.V., Oziraner A.S.
[1]   *Stability and stabilization of motion with respect to a part of variables.* Moscow: Nauka, 1987.

Samoilenko A.M., Perestyuk N.A.
[1]   *Impulsive differential equations.* Singapore: World Scientific, 1995.

Sanders J.A., Verhulst F.
[1]   *Averaging methods in nonlinear dynamical systems.* New York: Springer, 1985.

Šiljak D.D.
[1]   *Large-scale dynamic systems. Stability and structure.* New York: North-Holland, 1978.

Starzhinsky V.M.
[1]   *Applied methods of nonlinear oscillations.* Moscow: Nauka, 1977.

Stocker J.
[1]   *Nonlinear oscillations in mechanical and electrical systems.* Moscow: Inostr. Lit., 1953.

Utkin V.I.
[1]   Variable structure systems with sliding modes. *IEEE Trans. Autom. Contr.* **22** (1977) 212–222.

Vinograd R.E.
[1]   On the boundedness of solutions of regular systems of differential equations with small components. *Uspechi Mat. Nauk,* **8:1 (53)** (1953) 115–120.

Volosov V.M., Morgunov B.I.
[1]   *Averaging method in the theory of nonlinear oscillatory systems.* Moscow: Publishing House of Moscow Univ., 1971.

Vorotnikov V.I.
[1]   *Stability of dynamical systems with respect to a part of variables.* Moscow: Nauka, 1991.

Walter W.
[1]   *Differential and integral inequalities.* Berlin: Springer, 1970.

Xinzhi Liu, Shaw M.D.
[1]   Boundedness in terms of two measures for perturbed systems by generalized variation of parameters. *Communications in Applied Analysis,* **5** (4) (2001) 435–443.

Yakubovich V.A.
[1]   On the boundedness and global stability of solutions of some nonlinear differential equations. *Dokl. Acad. Nauk SSSR,* **121** (6) (1958) 984–986.

Yakubovich V.A., Starzinskii V.M.
[1]   *Parametric resonance in linear systems.* Moscow: Nauka, 1987.
[2]   *Linear differential equations with periodic coefficients.* New York: Wiley, 1975.

Yoshizawa T.
[1]   Lyapunov function and boundedness of solutions. *Mathematics, Coll. of Translations,* **9:5** (1963) 675–682.
[2]   *Stability theory by Liapunov's second method.* Tokyo: The Math. Soc. of Japan, 1966.

Zubov V.I.
[1]   *A.M. Lyapunov's methods and their application.* Leningrad: Publ. of Leningrad Univ., 1957.
[2]   *Mathematical methods of investigation of systems of automatic control.* Leningrad: "Mashinostroenie", 1974.
[3]   *Oscillations and waves.* Leningrad: Publ. of Leningrad Univ., 1989.

Index

M-matrix, 65

Abstract Cauchy problem, 180

Boundary of the cone, 14

Comparison functions of the same order of growth, 26

Comparison principle, 29

Condition $x = 0$
  asymptotically stable, 24
  exponentially stable, 26, 27
  uniformly asymptotically stable, 25
  uniformly stable, 24
  unstable, 28

Cone, 14

Cone-valued Lyapunov function, 31

Derivative Eulerian, 22

Dini derivative of the function v, 21–22

Domain $v_k > 0$, 92

Equilibrium
  μ-attracting, 34
  μ-stable, 34
  $x = 0$ (y, μ)-stable, 164
  asymptotically μ-stable, 34
  uniformly μ-stable, 34
  uniformly asymptotically μ-stable, 34

Equilibrium state
  completely μ-unstable, 104
  equistable, 86, 87
  exponentially μ-stable, 99, 100
  globally exponentially μ-stable, 102

globally uniformly asymptotically μ-stable , 98

uniformly asymptotically μ-stable, 96

Equilibrium state $x_k = 0$
  μ-stable, 91, 92
  μ-unstable, 93

Function
  x_i^{T}-decrescent, 106
  x_i^{T}-positive definite, 106
  decrescent, 20
    on $R_+ \times S$, 20
    with respect to y, 163
  globally
    decrescent on R_+, 21
    positive semidefinite, 18
  negative definite, 19, 20
  negative semidefinite
    in the neighborhood S, 18
  negative semidefinite (globally), 19
  positive definite, 19, 20
    globally, 20
    in the neighborhood S of the point $x = 0$, 19
    on $R_+ \times S$, 20
    with respect to y, 163
  positive semidefinite, 18, 19
  positive semidefinite
    globally, 19
    in the neighborhood S, 18
    on $R_+ \times S$, 19
  quasimonotone nondecrescent, 12
    with respect to the cone, 15
  radially unbounded, 21

strictly positive (negative) semi-definite, 19, 20

Generalized direct Lyapunov method, 182

Indirect control systems, 112
Inequalities
    Bihari, 7
    differential, 12
    Gronwall, 2
    integral, 16
Infinitisemal generator of semigroup, 180

Longitudinal motion of an aeroplane, 109
Lyapunov function for the system in Banach space, 183

Matrix
    strictly positive (negative) semi-definite, 19
    with the dominant main diagonal, 65
Measure ρ continuous with respect to the measure ρ_0, 40
    uniformly, 40
    asymptotically, 40
Motion
    $(\rho_0, \rho)\mu$-equibounded, 39
    μ-bounded, 50, 53, 56
        with respect to x_0 with respect to subvector, 59
        with respect to subvector, 57
    μ-bounded uniformly, 62, 65
        ultimately, 62, 65
        with respect to (t_0, x_0), 57
        with respect to (t_0, x_0) with respect to subvector, 60
        with respect to t_0 with respect to subvector, 57, 58
        with respect to x_0 with respect to subvector, 57
    uniformly $(\rho_0, \rho)\mu$-bounded, 39, 44

uniformly ultimately $(\rho_0, \rho)\mu$-bounded, 39, 45

Nonasymptotically stable subsystems, 151
Nonlinear semigroup, 180

Permissible pair $(Q(t), v)$, 184

Semigroup of bounded linear operators, 180
Solution
    μ-maximum, 47
    μ-minimum, 47
    $x = 0$ (y, μ)-stable, 165
    maximum, 12
    minimum, 12
Solution $x = 0$
    μ-stable, 123
    μ-unstable, 127, 146
    asymptotically μ-stable, 132
State $x = 0$ stable, 23
State of equilibrium, 33
Strengthened function, 41
Strengthened Lyapunov function, 55
System
    (ρ_0, ρ) μ-stable, 75, 77, 80–82
    μ-polystable, 105–107
    asymptotically $(\rho_0, \rho)\mu$-stable, 78
    uniformly (ρ_0, ρ) μ-stable, 85
    with an unstable free subsystem, 114
System of n oscillators, 171
System with weak linear connections, 69

Trivial solution stable, 181

Vector function, 185

Weakly connected Lurie–Postnikov equations, 67
Weakly connected mechanical system, 38
Weakly connected oscillatory system, 166

Milton Keynes UK
Ingram Content Group UK Ltd.
UKHW040101071024
449327UK00019B/721

9 780367 380632